LIBERAIDER!

THE MEN, MACHINES, AND MISSIONS

OF THE

461ST BOMB GROUP
49TH BOMB WING
FIFTEENTH AIR FORCE

DURING WORLD WAR II

CHARLES PARSONSON

LIBERAIDER!

THE MEN, MACHINES, AND MISSIONS

OF THE

461ST BOMB GROUP
49TH BOMB WING
FIFTEENTH AIR FORCE

DURING WORLD WAR II

BY
CHARLES PARSONSON

ISBN # 978-1-68593-268-8

Additional copies of this book are available from the Author's wife

Lori Parsonson, 55 Whistling Duck Drive, Bridgeville, DE 19933
l.parsonson@yahoo.com | 302.841.8962

Also Available on Amazon

WISE PUBLICATIONS
CUSTOM BOOK MANUFACTURING SINCE 1982
809 East Napoleon St., Sulphur, Louisiana 70663
337-527-8308 | wisepublications@yahoo.com
Visit our online Bookstore! **www.wisepublications.biz**

FORWARD

My father was an Air Force man. His dream was going to West Point and becoming an Army Air Force aviator. He applied, but was not selected by West Point and fell back on Plan B; he went to work at the Ford Plant along with his father and grandfather. However, he still dreamed of flying and put off his marriage because the Army didn't take married flight cadets.

Everything changed after the Japanese bombed Pearl Harbor and the Army dropped the restriction on married cadets. He applied to take the examination at the local cadet board, which he passed. Next was a five-day series of tests at the Army Air Force Cadet Training Center. One of the first things he was required to do was go up with an instructor who did every acrobatic maneuver he could think of. He would have walked off that field and gone home except my mother was there watching and he couldn't quit in front of her after making her wait so long. After the testing was completed, he became a pilot cadet and began the journey to becoming an Air Force pilot. He already knew he didn't want to become a fighter pilot.

Following getting his commission and wings he was assigned to a bombardier school flying students as they learned their trade. He later was assigned to B-24s, met his new crew, and headed to Italy where he joined the 764th Bomb Squadron, 461st Bomb Group. 49th Bomb Wing, Fifteenth Air Force. By the time the war ended he was a captain and squadron operations officer. He had flown 28 missions not counting double credit missions. And he had decided to stay in the Air Force, much to my mother's surprise who realized the house in Royal Oak, Michigan would have to wait.

I grew up in the Air Force and never really thought there was any other way to live. Mom and I would go out to the flight line to see dad return from wherever he went and I would get nervous when dad was late. Mom just said "he'll be back, he always comes back". I grew up all over the world, including seven years in other countries and twelve schools in twelve years, graduating from high school in Munich, Germany. I was a true Air Force brat who liked my civilian relatives but didn't really know them. I don't think they understood me any more than I understood them.

As might be expected I intended to become an Air Force officer and pilot. There was no plan B. I took the Air Force Officer Qualifying Test and scored 99th percentile, The pilot test was 97th percentile. Somewhere along the way someone neglected to tell me Air Force pilots couldn't wear glasses and it was time for Plan B. I decided not to make the Air Force my career, joined the army and went to Vietnam for 20 months. I went home, put the uniform away, and went about the business of jobs and family. The military became a dream once lived. Dad passed away and a few years later mom followed. They're together now at Arlington National Cemetery,

A few years before dad died, he mentioned that he had a favorite plane during the war. It always brought him back no matter how damaged it was. The one thing he wanted to know was what happened to it. I found he had flown it back to Hunter Field, parked it and walked away. It had been declared excess the next day and sent to Walnut Ridge Arkansas where it was melted down for scrap. Neither of us particularly liked its fate, but it did bring us closer together. I started studying the 461st Bomb Group and eventually became the Group Historian. I met other veterans and other families, many of which had no idea what dad or granddad did in the war.

As the past Historian for the 461st Bombardment Group Association, I have had the honor of working with veterans and their families who are searching for information about aircraft, missions, former friends, personnel files, or various other documents. Many veterans want to share their stories. Children and grandchildren want to know what dad/grandpa did in the war. Families made pilgrimages to where a family member gave his life. Some had misconceptions while others had no idea how the 461st fit into the Fifteenth Air Force. Each person, whatever the question, was a joy to assist. Later, with the number of veterans declining, other groups joined the 461st reunions, opening up an entirely new group of people asking the same questions. One Saturday I received

an email from a man named Frank asking for help. He had been looking for his records for years but nobody could help him. Another man told him if I couldn't do it nobody could. He didn't know his group, crew, missions, planes, anything. Just to put a little pressure on me, he said he was using his daughter's computer who would be leaving the next day. I had no idea where to start so I began by looking through group rosters. It was on the third group that I found him. This was only the beginning of a lucky night. I had all his groups records sleeping peacefully in my computer. I found Frank along with his mission loading lists which included his entire crew along with their rank and position, his planes serial number and the target. Before I went to bed I had everything he was looking for along with a photo of each of the aircraft he flew. I called him the next morning and told him the news. What followed was hours of questions and answers. He finally told me he had to go because he was being called to dinner. I let him know all the information would be on his computer before he got back from dinner. His only question was "can I call you sometimes?" Yes, you bet.

Any attempt to write a book about a combat group inevitably ends up being predominantly about combat, and much of this effort will include that subject. However, most members of the group did not engage in combat. That did not make their contribution to defeating the enemy any less. Given the lack of documents after so many years, every attempt has been made to identify all members of the group, their units, and their jobs, and to give them the proper recognition they deserve.

Finally, much as we would like to believe the 461st won the war, it was in reality part of a much larger organization. This work is also an attempt to establish the relationships between the 461st Bombardment Group, Fifteenth Air Force, Fifteenth Air Force Service Command, and Army Air Force Service Command/Mediterranean Theater of Operations (AAFSC/MTO). To be clear, this is not a work for serious researchers. It is, hopefully, what kids, grandkids, and families would find helpful in understanding what dad did in the war.

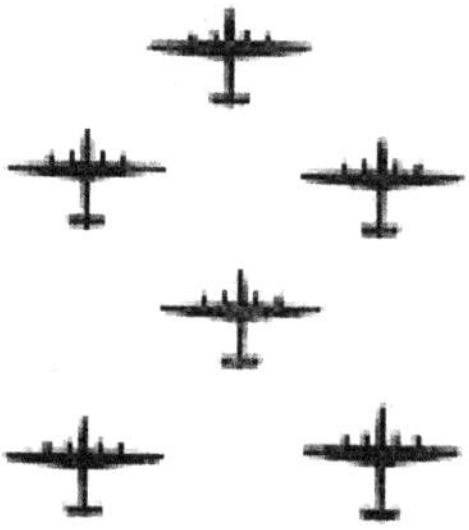

TABLE OF CONTENTS

Thank you to Jonathan Tudor for his time, effort,
and dedication in bringing Chuck's book to publication.

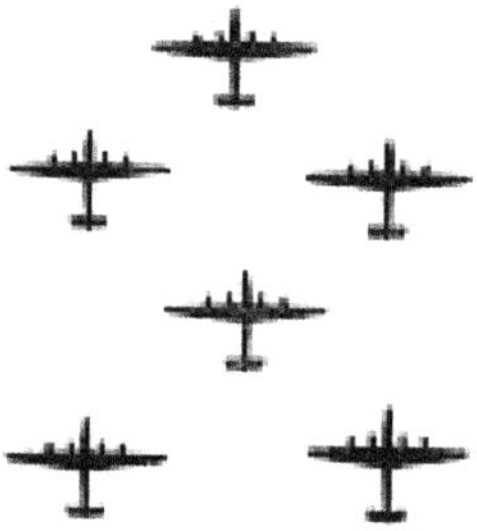

Thank you to Jonathan Tudor for his time, effort,
and dedication in bringing Chuck's book to publication.

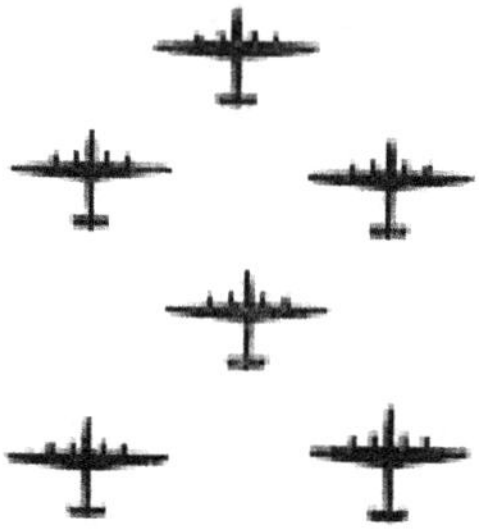

(1)

EARLY HISTORY

FIFTEENTH AIR FORCE – GOALS & STRATEGIES / THE SECOND AIR FORCE AND CREATION OF THE 461ST

FIFTEENTH AIR FORCE–
Goals and Strategies

On 1 November 1943 Headquarters, Fifteenth Air Force sent an advance party to Bari, Italy, to negotiate for a headquarters complex. A few days later the Commanding General, Twelfth Air Force authorized the Fifteenth Air Force to relocate their headquarters to Bari, and the Commanding General, Fifteenth Air Force authorized the advance party to complete the arrangements for the new facility. Instructions were issued to begin the move on 30 November 1943. Almost the entire staff plus equipment, except motor vehicles, were relocated by Air Transport Command, moving 494 officers and men in 89 flights over just 4 days. The remaining 50 officers and men were relocated between 4 December and 18 December 1943 along with another 28,000 pounds of equipment. The old headquarters in Tunis and the new headquarters in Bari were opened simultaneously on 1 December 1943.

*   *   *   *   *

Initially, the reorganization of the Twelfth Air Force and creation of the Fifteenth Air Force involved the transfer of all Twelfth Air Force tactical units, comprised of two heavy bomb wings (four B-17 and two B-24 bomb groups), two medium bomb wings (three B-26 and two B-25 bomb groups), and four fighter groups.

When these units arrived at the Fifteenth Air Force, one heavy bomb wing was eliminated and the two B-24 bomb groups (98th and 376th) from that wing were assigned to the 5th Bomb Wing, joining the four B-17 groups (2nd, 97th, 99th, and 301st) and two fighter groups (14th and 325th). Three B-26 bomb groups and one fighter group made up the 42nd Bomb Wing, while the two B-25 bomb groups and one fighter group formed the 47th Bomb Wing.

That reorganization lasted only three days when the two B-25 groups were transferred back to the Twelfth Air Force, leaving one fighter group in the 47th Bomb Wing. The 98th and 376th Bomb Groups (B-24), just assigned to the 5th Bomb Wing, were reassigned to the 47th Bomb Wing, changing it from a medium bomb wing to a heavy bomb wing.

More changes occurred on 1 January 1944 when the three B-26 bomb groups were also reassigned to the Twelfth Air Force. This left the Fifteenth Air Force with six heavy bomb groups and four fighter groups. The goal was to reach 21 heavy bomb groups, 7 fighter groups, and one reconnaissance group by March 1944. That goal turned out to be overly optimistic. Although the 49th Bomb Wing was assigned to the Fifteenth on 1 January 1944, it did not become fully operational until 9 April 1944. By then the 451st and 461st had already flown their first missions.

* * * * *

The Casablanca Conference, held at the Anfa Hotel in Casablanca, French Morocco from 14 to 24 January 1943, set the Allied strategy for the next phase of the European war effort. Attending were President Franklin D. Roosevelt, British Prime Minister Winston Churchill, and the Combined Chiefs of Staff. The agenda included tactical procedures, allocation of resources, and diplomatic policy. It also produced the "Casablanca Declaration," demanding the unconditional surrender of all Axis powers, and the "Combined Bomber Offensive Plan." The joint British-American plan laid out the strategy to destroy German transportation and industry, with goals to:

1. Cripple or destroy German aircraft fighter strength,
2. Destroy petroleum, oil, and lubrication (POL) facilities,
3. Destroy railyards and other transportation targets, and
4. Support ground forces.

Conference of the combined Chiefs of Staff held 4 December 1943 at the Mena House Hotel in Cairo, Egypt

President Franklin D. Roosevelt meets with Prime Minister Winston Churchill at the Casablanca Conference. Behind them are Lt. Henry (Hap) Arnold, Admiral King, General Marshall, Admiral Sir Dudley Pound, General Sir Allen Brook, and Air Marshall Sir Charles Portal.

General Marshall and Lt. General Henry (Hap) Arnold at the Casablanca Conference.

As mentioned above, the Fifteenth Air Force had four primary objectives when offensive operations began, and those objectives remained throughout the war:

First, to destroy the German Air Force in the air and on the ground. It was believed that unless the GAF was neutralized, the Allies could not sustain a bombing campaign without unsustainable losses. This included destroying German airframe and component production facilities, and the ball-bearing industry.

Early in the Luftwaffe's history, production for all aircraft types was only 500–600 per month (1939–42). In January 1942, General Milch took over supply, increasing production to about 1,000 per month by mid-1943, mostly fighters. Despite advice to prioritize fighters, bomber production continued until March 1944. Heavy damage from Allied bombers eventually forced Germany to emphasize fighter production. Output rose from 1,377 fighters in March 1944 to 3,081 in September, but production steadily decreased from there until the end of the war. Hitler and Goering had demanded, as late as July 1944, the continued production of some bombers. It was also at this time that Hitler ordered the development of the jet-powered Me-262 as a bomber, a decision that delayed its development as a fighter. The GAF did try switching to exclusively producing Me-262s fighters in March 1945, but it was too late. The war had already been lost.

The air campaign named POINTBLANK was initiated on 11 January 1944, although the Fifteenth Air Force had already begun counter-air operations from Tunisia and Italy in late 1943. POINTBLANK, based on the Combined Bomber Offensive Plan, required attacks on German aircraft factories, ball-bearing plants, and related industries. RAF Bomber Command struck by night, while the Eighth and Fifteenth Air Forces carried out daylight precision bombing.

The Luftwaffe responded by dispersing production, increasing fighter output, and employing new tactics such as head-on bomber attacks and night interceptions. Losses of aircraft and experienced pilots, combined with Allied long-range escorts like the P-51, steadily weakened the GAF.

The Fifteenth Air Force attacked major fighter production plants at Wiener Neustadt, Regensburg, and Augsburg; jet factories at Schwechat, Budapest, Oberpfaffenhofen, and Friedrichshafen; and ball-bearing plants at Steyr, Klosterle, Schweinfurt, Turin, and Villa Perosa. The Schweinfurt raids inflicted heavy damage but came at great cost, particularly the second raid ("Black Thursday"), when 60 B-17s were lost.

Although POINTBLANK guided strategy, directives from USSTAF and the Mediterranean Allied Air Force (MAAF) occasionally shifted emphasis. For example, during the Anzio landings, counter-air operations were suspended to focus on rail lines. In June 1944 oil became the primary objective, and in January 1945 jet fighter production was given top priority alongside oil. Strategic operations ended 16 April 1945.

* * * * *

Second, to cripple the German war machine by denying it oil.

Within 1,400 miles of the Foggia complex were thirty-one plants producing nearly half of Germany's crude and synthetic oil. Another twenty-one plants within 800 miles produced an additional thirty

---

Pointblank directive - Wikipedia
POINTBLANK DIRECTIVE | American Air Museum
Operation POINTBLANK: A Tale of
D-Day: Operation Pointblank - Historic UK

percent, for a combined 11,825,000 tons annually. Synthetic plants were given priority over natural refineries due to their operational flexibility.

Key objectives for the Fifteenth Air Force were:

1. Destroy Romanian oil refineries, especially Astra Romana, Concordia Vega, and Romana Americana,
2. Destroy synthetic plants at Brux, Blechhammer North, Blechhammer South, and Odertal,
3. Eliminate natural refineries at Budapest, Almasfuzito, Vienna (Lobau and Florisdorf), Kragan, Pardubice, Bratislava, and Kolin,
4. Keep northern Italian and southern French refineries non-operational.

Crude fields in Romania, Hungary, and Austria added 6.8 million tons annually. Repeated attacks on these facilities severely disrupted German fuel supplies and were considered decisive to the Allied air campaign.

* * * * *

Third, to support the battle on the Italian mainland through communications interdiction in Italy, Austria, and along the Brenner Pass.

The Fifteenth Air Force flew missions against road, rail, and shipping targets across eleven countries, dropping more tonnage on communications than on all other target categories combined. About one-third of all aircraft losses came from these missions. Railroads, especially marshalling yards, were frequent primary or secondary targets, and over 90% of interdiction missions struck railroad facilities. These attacks disrupted German command and control, supply distribution, and morale.

* * * * *

Fourth, to coordinate with ground units and weaken German positions in Italy and the Balkans.

Although not its highest priority, the Fifteenth flew 11,281 sorties in direct support of ground forces, dropping 25,374 tons of bombs. Attacks on airfields and communications not only weakened the GAF but also hindered German reinforcements and unit coordination. Support operations during Anzio, Cassino, and in the Balkans demonstrate how counter-air and counter-oil missions often directly benefited ground campaigns.

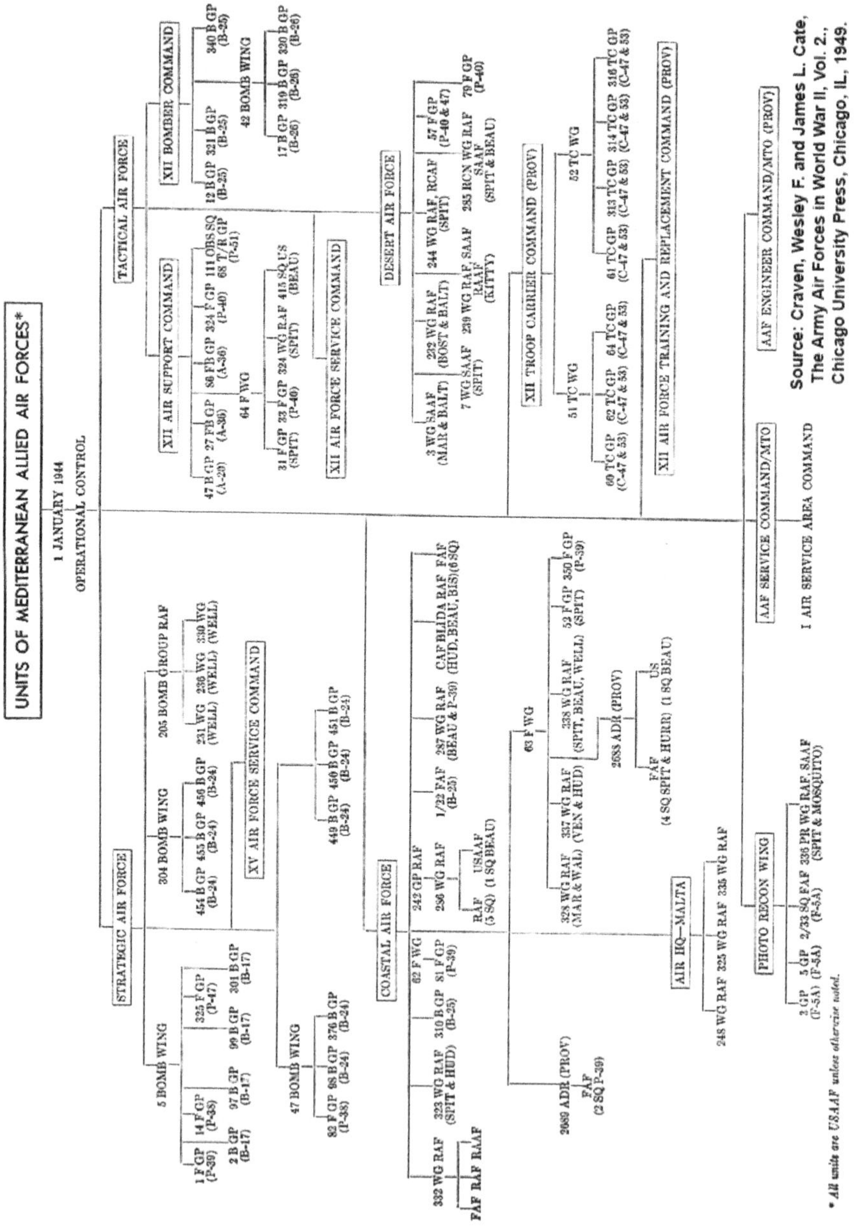

Source: Craven, Wesley F. and James L. Cate, The Army Air Forces in World War II, Vol. 2., Chicago University Press, Chicago, IL, 1949.

They could not call for reinforcements or coordinate defensive strategies effectively, making them easier targets[1]. The constant disruption of communication networks also had a psychological impact on German troops. The uncertainty and isolation caused by poor communication contributed to lower morale and increased stress among soldiers[1]. The Germans attempted to mitigate these effects by using alternative communication methods, such as couriers and signal flares, and by improving their encryption and radio technologies. However, these measures were often insufficient to fully counter the impact of Allied bombing campaigns[1].

Overall, the disruption of communication networks significantly weakened the German military's ability to coordinate and execute effective operations, contributing to the overall success of the Allied campaigns.

* * * * *

Fourth, to coordinate with ground units and to weaken German positions in Italy and in the Balkans.

While not the same priority of the air and oil campaigns, the Fifteenth Air Force would be called on to support advancing Allied ground forces where it made sense and with the approval of the Commanding General, USSTAF and the Commanding General of the Air Force. While many of the contributions made by the Fifteenth Air Force to the various ground campaigns may appear to be more related to counter-air or counter-oil operations, it actually demonstrates the effect one operation has on another. For example, bombing a German airfield could destroy a number of enemy aircraft, but also makes it much more difficult for the GAF to land reinforcements during the coming ground offensive. Destroying German communications networks make coordination of German army units trying to defend against advancing Allied armies impossible. Total Fifteenth Air Force operations in support of ground units amounted to 11,281 sorties, which dropped 25,374 tons of bombs on military targets.

THE SECOND AIR FORCE AND THE CREATION OF THE 461ST

IMMEDIATELY FOLLOWING THE PEARL HARBOR attack on 7 December 1941, 2nd Air Force was placed under Western Defense Command and tasked with organizing air defense for the northwest Pacific Ocean coastline of the United States and with antisubmarine patrols along coastal areas until October 1942. In late January 1942, elements of the B-25 Mitchell-equipped 17th Bombardment Group at Pendleton Field, Oregon were reassigned to Columbia Army Air Base, South Carolina ostensibly to fly antisubmarine patrols off the southeast coast of the United States, but in actuality came to prepare for the Doolittle Raid against Japan.

On 5 January 1942, it was placed under Air Force Combat Command. The mission of Second Air Force became supporting the Army Air Forces Training Command's mission of training units, crews, and individuals for bombardment, fighter, and reconnaissance operations. After personnel graduated from AAFTC flight schools; navigator training; flexible gunnery schools and various technical schools, Second Air Force organized the personnel, aircraft and equipment into combat groups and squadrons. The newly formed units received secondary training prior to their assignment to the deployed combat air forces in the various overseas theaters. Second Air Force became the training organization of B-17 Flying Fortress and B-24 Liberator heavy bombardment groups. Nearly all new heavy bomb groups organized after Pearl Harbor were organized and trained at Second Air Force Bases, by II Bomber Command operational training units (OTU) were then deployed to combat commands around the world.

The 461st Bombardment Group (H) was activated on 1 July 1943 by General Order 78, dated 29 May 1943. It was organized under the Second Air Force at Wendover Field, Utah as an organizational Training

Unit for heavy bombardment crews. Most of the original personnel who established the 461st, and who would serve as their training cadre came from the Second Air Force, 29th Bomb Group.

AIR CREW TRAINING

As has been mentioned earlier, a large amount of training took place before personnel arrived at the 461st. Anyone interested in cadet training was required to pass an examination at the local cadet board. If successful, the next step would be a series of tests at the Army Air Force Cadet Classification Center, a process taking three to five weeks. Based on the results of those exams candidates could be placed in either pilot, bombardier, or navigator programs.

Preflight School – Those selected as future pilots would undergo an intensive nine week pre-flight program in physical training, military and academic work. Academic subjects include mathematics, physics, military law, citizenship, national policy, organization of the United States government, and the Army Air Force.

The PT-17 Stearman was primarily used as a basic trainer aircraft for the United States Army Air Forces (USAAF), Navy, and Marine Corps. It was designed to train new pilots in the fundamentals of flight, including takeoff, landing, and basic maneuvers.

Primary School – The next nine weeks are spent at one of the Army Air Forces Training Center's elementary flying schools, such as Cal-Aero in Ontario, California. These schools, located at civilian airports, were under contract to the Army Air Force. Primary provides a basic foundation in flying and the opportunity to move up to a more demanding program. Cadets would have soloed, and performed aerobatics including loops and snap rolls. They could put their ships in flat spins and bring it out again. Many going into Primary fail to achieve these goals and wash out due to a lack of "inherent flying ability". However, thousands of cadets do successfully complete Primary every nine weeks and move on to Basic Training.

The North American BT-9 was the United States Army Air Corps (USAAC) designation for a low-wing single engine monoplane primary trainer aircraft that served before and during World War II.

Basic Training – After completing Primary at Cal-Aero, Douglas, or some other school, the cadet is ready for Basic. Contrary to how it sounds, Basic involves more challenging equipment, aircraft, and flying techniques. Aircraft will have greater weight, speed and horsepower. He would spend approximately 70 hours in the air, learn to fly at night, by instruments, in formation, and cross country from one point to another. It was also the time to determine if he went to single-engine or multi-engine advanced flying school.

The Curtiss-Wright AT-9 was an American twin-engine advanced trainer aircraft used by the United States during World War II to bridge the gap between single-engine trainers and twin-engine combat aircraft.

Advanced Training – Cadets advancing to multi-engine school received another 70 hours in the air during the next nine weeks. Unlike their single-engine peers, they did not practice combat maneuvers or aerial gunnery, but did learn the art of flying a multi-engine aircraft in formation and increased their skills in flying on instruments at night. Their first 10 flight hours involve the transition to faster and heavier ships with retractable landing gear, hydraulic flaps, and constant speed props. They also spend time in Link trainers learning what will be practiced later in the air. Another 10 hours of Advanced Training involved flexible gunnery training at another school. At the end of the nine-week program cadets received their wings and commission as Second Lieutenants. Before pilots begin to train in units, they must learn to fly the aircraft they're going to fly in combat. B-24 and B-17 pilots got 105 hours 4-engine flying time plus additional ground school training in a 10-week post-graduate transition course.

Finally, members of a crew report to groups at one of the four domestic air forces where they undergo a 90-day training program divided into three parts. First, they increase their individual skills, learn to be part of a team, and become familiar with equipment and techniques. Second comes formation flying. Third, they move to a training area that approximates a combat zone to practice the skills they've learned.

Navigators go from the Classification Center to a nine-week preflight program that is very similar to the pilot program and follows the same three categories although the academic component places more emphasis on mathematics. They are also taught Morse code, flags of all nations, physics, naval forces, meteorology, photography, maps and charts, communications, cryptography, and target identification.

Following that was a 15 week advanced navigator school followed by flexible gunnery school for 5 weeks. Although the pilot, co-pilot, navigator, and bombardier have distinct duties, there may come a time when they would be called upon to man a machine gun to defend the ship. After all his training the last step is commissioning and assignment to a crew.

Bombardiers, like navigators, move from the classification center to a nine-week program at one of the many bombardier schools around the country. His studies stressed bombardment aviation and observation from the aircraft. Most other programs were the same as pilots and navigators.

All enlisted crewmen are trained as **aerial gunners**. However, only career-gunners are exclusively gunners. All others are trained in a technical specialty before attending flex gunnery school. At the end of basic training, men chosen to train as **Gunners** are allowed to enter a six-week flexible gunnery school that began with firing pellet guns at indoor targets. Next came skeet shooting with shotguns. Finally, .30 and .50 caliber machine guns were used from moving trucks at tow targets. Also required in addition to weapons training was ballistics, turret operation and maintenance, gun repair, sea and land recognition, aircraft recognition, range estimation, Morse code, and the theory of ballistics. Individuals selected as **airplane armorer-gunners** take a twenty-week course in the operation and maintenance of aircraft armament. **Airplane mechanic-gunners** take a 27 week training course in airplane inspection and maintenance of airplane. **Radio operator mechanic-gunners** get 20 weeks of combat crew radio operation and repair.

The final step is combat crew school, or Transition training where pilots are matched up with the other nine men that will form their crews. Each man is proficient in his position. After transition training, he will be part of a crew who will receive orders to report to the 461st Bomb Group.

SUNNY ITALY:
Movement of the Ground Echelon

THE 461ST BOMBARDMENT GROUP would begin its move to Italy and the Fifteenth Air Force on 1 January 1944. The training had been completed and it was time to face the enemy. The first train, carrying the equipment and the personnel of the 765th Bomb Squadron left Hammer Field for Camp Patrick Henry, VA on 31 December 1944. The next morning the Headquarters Detachment and the 764th Bomb Squadron left aboard the second train. The final train, carrying the 766th and 767th Bomb Squadrons to Patrick Henry left later that day. Six days later, at 2100 o'clock on 6 January 1944 the first train arrived at Camp Patrick Henry, twelve miles from Hampton Roads, VA, the Point of Embarkation. The second train arrived three hours later and the last train arrived early the following morning.

The group would remain at Patrick Henry for five days where they were issued clothing, equipment, and arms in accordance with the directive for overseas travel. All personnel were also given physicals. Only one person failed the physical and was lost to the group. Two others went AWOL during this period.

On 12 January 1944, the group left Camp Patrick Henry for Hampton Roads by train. Upon arrival they went directly to the following liberty ships:

S/S Archbishop Lamy	*HQ Detachment and 764th Squadron*
S/S George S. Hanley	*765th Squadron*
S/S William Rawle	*766th Squadron*
S/S John Jay	*767th Squadron*

After boarding, the ships moved out into the bay to wait for other ships to load and join the convoy. The convoy, made up of 75 ships along with 25 escort vessels, was completed by the evening of 13 January and sailed the following morning.

On 31 January, the convoy entered the Mediterranean Sea where British escort ships replaced the American escorts. The following day the convoy was attacked by German aircraft, which according to unofficial reports, sank one ship and damaged another, while losing one aircraft in the attack. On 4 February three of the four liberty ships dropped anchor in the harbor at Tunis and lay off shore for six days. They finally left Tunis on 10 February, arriving at the harbor of Naples two days later. The group promptly went ashore and moved to Bagnoli, seven miles from Naples. Trains arrived at Bagnoli on the 16th, 17th, and 18th to take the group to Cerignola, a 36 hour trip by "40 and 8" cars. Once at Cerignola they boarded trucks for a 2½ hour trip to Venosa, the intended new home of the 461st. However, the field was totally unprepared for combat operations. Partly due to the late arrival of 49th Wing Headquarters the fields assigned to their groups were not ready when they arrived. The 460th Bomb Group, originally assigned to the 49th Wing, was assigned to Torretta Field. Since the 49th Wing had not arrived, they were reassigned to the 55th Wing and moved to Spinnazola. The 461st arrived in February 1944 prior to the arrival of the 49th Wing so it was also assigned to the 55th Wing and assigned to Castellucio. When ready for training missions in March it was still unable to occupy its assigned field (Venosa) so the Commander of the 55th Wing, Colonel George Acheson, assigned the group to the now ready Torretta Field, eight miles south of Cerignola. On 14 March 1944 the 461st was reassigned to the 49th Wing. In the meantime, the 484th was assigned to Palazzo in the 55th Wing area, which was abandoned in March 1944 due to difficulties in construction leaving the 55th Wing with no available airdromes. Spinazzola had been occupied by the 460th (formerly 49th Bomb Wing), and Pantanella and Venosa were still under construction. On 14 March 1944 the 484th was reassigned to the 49th Wing and moved to share an enlarged field with the 461st at Torretta.

* * * * *

The Liberty Ship, the S/S George S. Hanley, which was carrying the 765th Squadron, did not put in at Tunis like the rest of the group. Instead they were taken to Augusta, Sicily. After a week there, they were transported to Naples and trucked to Bagnola. The following day they boarded a train only to be taken to Barletta. From there they were hauled by truck all the way to Torretta.

MOVEMENT OF THE AIR ECHELON

The air echelon of the 461st Bombardment Group was scheduled to leave the United States in mid-January 1944 with arrival in Italy mid-February. It was decided that the first squadron to complete their minimum training requirements would receive their full complement of aircraft and would be the first to leave for the processing field. That honor was earned by the 766th.

STATION	766 SQUADRON	767 SQUADRON	764 SQUADRON	765 SQUADRON
Hamilton Field, California	Jan. 14	Jan. 17	Jan. 19	Jan. 20
AAB, Palm Springs, Calif.	17	21		25
Sky Harbor Airport, Phoenix, Ariz.			21	
AAB, Midland, Texas	18	22	22	27
AAB, Memphis, Tenn.	20	23	23	28
Morrison Field, W. Palm Beach, Fla.	21	24	24	29
Waller Field, Trinidad, British, W.I.	25	27	27	Feb. 1
AAB, Belen, Brazil	26	28	28	2
AAB, Fortaleza, Brazil	27	29	29	
AAB, Rufisque, (Dakar) Fr. W. Africa	29	31	31	5
AAB, Marrakech, French Morocco	30	Feb. 1	Feb. 1	6
AAB, Telegma, Algiers	31			
AAB, Djeseida, Tunisia	Feb. 2	2	2	
AAB, Oudna, Tunesia	3	3	3	8
AAB, Torretta, Italy	22	22	23	23

MORRISON FIELD

The last stop in the United States was Morrison Field at West Palm Beach, Florida. It was this station that assured crews had aircraft equipped for an Atlantic flight, and the information and training to make that flight. They made sure that all crew members were equipped with route data, maps, navigational equipment, secret codes, and other necessary items.

When a crew arrived they were directed to group operations to schedule their briefings, normally held three times daily but at least 24 hours prior to their scheduled departure. The briefing was designed to review the route folder with the pilot and navigator, give the pilot, co-pilot, and navigator an opportunity to review the folder and ask any questions, allow the pilot and crew time to ask questions of experienced personnel at the base, and to allow the navigator time to calibrate his instruments. The latest weather update would wait until just before takeoff. Individual packets were used for the pilots and navigator and both packets were discussed in great detail. The navigator had a one-on-one discussion with the senior Morrison navigator and the plane was not permitted to attempt the Atlantic crossing until the briefing navigator was convinced the ship's navigator was equipped and ready to make the crossing. No night flights were permitted without a navigator.

Maps were also provided for the entire route which covered all four of the methods of navigation: pilotage, dead reckoning, celestial, and radio.

The pilot would have been given additional information for emergency situations that included routes to alternate airfields. It also included descriptions of the fields, radio stations and frequencies, housing and eating facilities, and landmarks to aid navigation.

* * * * *

The 461st would send 70 combat crews and 62 B-24s to Torretta Field without the loss of a single life. However, one ship, 42-52379 of the 764th Bomb Squadron would be lost at Oudna, Tunisia when the co-pilot, while taxiing the plane, accidentally pulled up the landing gear instead of the flaps. Upon arrival they were ordered to transfer 10 additional ships to the 47th Bomb Wing, leaving just 51 aircraft for the 70 crews. It did not take long to realize nobody would have their own ships.

More problematic would be the decision to transport 248 officers and men as passengers aboard the 62 ships in addition to the 620 crewmen. Since the movement was under the control of Air Transport Command, the Group was told upon arrival at Morrison Field that passengers were no longer allowed to cross the Atlantic on heavy bombers. The Group was ordered to offload all 248 passengers. After remaining a minimum of two weeks at Morrison Field, many were eventually transported to New York, where they made travel arrangements to Oudna, Tunisia. 130 enlisted personnel went by train to Camp Patrick Henry, Virginia. They finally left Hampton Roads on the fast, unescorted ship S/S General

Hann to Casablanca and Tunis. After a week in Tunis they were sent to Algiers where they were finally located by the 461st. They were transported by the Group's bombers to Torretta Field, arriving on 30 April 1944. It had taken 3½ months to make the trip.

TORRETTA FIELD

If Venosa was in no condition to support combat operations, Torretta had not been that much better. However, the Engineer's Report showed a number of improvements shortly after offensive operations began:

ENGINEER'S REPORT ON TORRETTA FIELD

Dated October 15, 1943, Revised April 14, 1944

Compiled by S-3 Section

21st Engineers

- **Name and Location:** Torretta, Cerignola
- **Geographic Location:** 41 10' 59" N Latitude, 15 45' 52" E Longitude
- **Map Reference:** 1:100,000 sheet 175mCerignola
- **Magnetic Variation:** 2' 51W January 1943 Annual change 3' East
- **Landmarks:** ¼ mile West of Mass Torretta. 1 mile South Cerignola, Candela Road. 8.6 miles Southwest of Cerignola.
- **Dimensions:** 8000 feet North/South by 8000 feet East/West landing ground.
- **Runway:** 150 by 6000 feet, bearing North 21 degrees West. Taxiways, Runway, and Hardstands are gravel topped. 112 H/S Surface. All in cultivation. Black loam soil. 18 inches top soil over chalice subsoil. Runways, Taxiways, and hardstands are of clay gravel.
- **Markings:** No markers on the runway. No circle and name at the time of reconnaissance.
- **Lighting:** None
- **Remarks:** Field operational. Hardstands and drainage system still under construction.

- **Obstructions:** Secondary power line on East side of field 30 feet high.
- **Hangers:** None
- **Repair facilities:** Service squadron units
- **Fuel and oil:** 1 1000 barrel storage tank for 100 octane fuel located ½ mile north of 3 outlets located at north end of the field. Fed by 4 inch pipeline.
- **Communications:** Telephone to Cerignola with connections to Bari and Foggia.
- **Radio communications:** Unknown.
- **Weather service:** Portable field unit.
- **Transportation:** No regular transportation. Military only from areas south side of field.
- **Obstructions:** Secondary power line on East side of field 30 feet high.
- **Hangers:** None
- **Repair facilities:** Service squadron units
- **Fuel and oil:** 1 1000 barrel storage tank for 100 octane fuel located ½ mile north of 3 outlets located at north end of the field. Fed by 4 inch pipeline.
- **Communications:** Telephone to Cerignola with connections to Bari and Foggia.
- **Radio communications:** Unknown.
- **Nearest seaport:** Manfredonia port of entry for 100 octane fuel. Bulk of supply shipments to Bari. 5000 feet of floor space. Capacity 14 Liberty ships.
- **Defensive Installations:** Units of British 40 millimeter anti-aircraft guns.
- **Civil Operations:** None. Tactical operations only.
- **Hospitals and First Aid:** Field medical units. Nearest general field general hospital in Cerignola.
- **Railroads:** Nearest railroad and sidings at Cerignola station 3.8 miles north of Cerignola. Three sidings 1800 feel each. Two sidings 900 feet each.

- **Highways:** Macadam highway 16 feet wide Cerignola to Bari and Foggia. 14 feet wide gravel top from Cerignola to field. 0.7 Mile gravel access road of Candela Road to field.
- **Waterways:** None
- **Electric Power:** No electric power service, except individual field units.

LOCATION OF FACILITIES:

- **Water Point:** Aqueduct way station at Candela 0.7 miles northeast of field. Four outlets, 1000 gallon storage tank and pump, used in conjunction with San Giovanni.
- **Fuel tanks** and three outlets at North end of field.
- **Radio antennas:** transmitter and receiver one mile north of field. British field units
- **Expansion ability:** Expansion used to the fullest extent.
- **Construction capability:** Cross wind runway not feasible.
- **Air Units:** British squadrons of Wellington bombers.
- **Weather service:** Portable field unit.
- **Transportation:** No regular transportation. Military only from areas south side of field.
- **Prevailing winds:** Unknown
- **Precipitation:** Unknown.
- **Visibility:** Fog reported less than 20 days a year during which time visibility is not greater than 2 ½ miles.
- **Operated by:** British R.A.F.
- **Fire Fighting Equipment:** Unknown
- **Additional Information:** This field is under the compete operation of the British RAF.
- **Mean monthly precipitation** in inches and temperature in degrees Fahrenheit.
- **Rainfall in inches:** Jan: 2.1, Feb: 1.1, March: 1.3, April, 1.4, May: 1.8, June: 1.1, July: 0.6, August: 1.2, September: 1.4 October: 2.1, November: 2, December:?
- **Days of rain per month:** January: 8, February: 7, March: 7, April: 9, May: 7, June: 6, July: 2, August: 4, September: 5, October: 8, November: 2, December: 7

- **Temperature:** January: 44, February: 49, March: 45, April: 60, May: 69, June: 76, August: 79, September: 71, October: 61, November: 52, December: 47
- **Additional Field Information** from Data Compiled April 14, 1944.

LANDING AREA:

- **Runways:** 150 feet by 6000 feet surfaced with gravel. 1000 feet of pierced steel planking on south end of dual runways and 500 feet a north end.
- **Control tower** and windsock located east end side and at midpoint of right runway.
- **Accommodations:** Tent areas, and administration buildings approximately 3000 feet south of the runway.
- **Transportation:** GI bus runs hourly to Cerignola.
- **Hangers:** None
- **Repair Facilities:** Service squadrons
- **Communications:** Telephone and administration Teletype at field. VHF, ADC being installed. No HF or DF.
- **Fire fighting:** Mobile Units.
- **Electric Power:** Portable field units and part time Italian power source 160-165 volts, 260 volts 3 phase 50 cycles.
- **Weather service:** Station at Cerignola
- **Prevailing winds:** Northwest Southeast
- **Markings:** Circle "TOR" at Northeast corner of runway. Red and white markers along the runway.
- **Lighting:** Runway lights for emergency landings.

AIRPORT DESCRIPTION

- **Dispersal area:** 128 hardstands gravel and pierced steel planking. Taxiways surfaced with clay gravel.
- **Facilities:** Unknown
- **Hospital and First Aid:** Dispensaries, ambulances at field. Hospital at Cerignola.
- **Air units:** Two heavy bombers groups, United States Army Air Forces Company 'H', 3rd Battalion, 21st Engineer Aviation Regiment, started work on the field late in November 1943 and remained at work with all platoons. They were relieved by the 2nd Battalion,

21st Engineer Aviation Regiment on 1 June 1944. They built a gravel runway measuring 150' by 6000' and laid 500' of Pierced Steel Planking (PSP) at each end of the runway. They widened 17,000 linear feet of compacted gravel taxiway 15 feet and built 50 heavy bomber hard stands.

They later laid 6,000 pieces of PCP on 800' at the south end of the R/W as an experiment at the insistence of the Air Corps. However, the planking rutted up due to foundation failure after a few days of heavy use.

The runway also had a serious drainage problem during heavy rains from January through March due to the poor soil conditions. Since it was too difficult to put down gravel during rain because it held water, the runway was opened to a depth of 30 inches, aired out and allowed to dry. The gravel was recompacted to a depth of 6 to 8 inches. Since the result was an extremely flat surface which caused water to stand on the surface. That problem was resolved by installing a drainage field.

Company 'F', 2nd Battalion, 21st Engineer Aviation Regiment resumed responsibility for the field on 29 May 1944. Their first act was the construction of a Butler Combat Hanger. Next, 100' by 1000' of PSP was laid on both ends of the west runway. Drainage ditches were dug and the runway and taxiway were maintained. In July 1944 maintenance was taken over by the 1898th Engineer Aviation Battalion. A detachment of 'D' Company, 21st Engineer Aviation Regiment was also assigned to assist the 1898th by constructing an all-weather gravel type 100' by 1000' runway on the site of the current dry weather runway. On 21 December 1944, Company 'F', 2nd Battalion, 21st Engineer Aviation Regiment again resumed responsibility for Torretta Field. During January and March 1945 they built a new gravel lift on the east runway to replace the original surface which had too much clay. Company F would be responsible for the field until the end of the war.

Torretta Field – The 484th Bomb Group occupied everything to the right of the runways, while the 461st Bomb Group was to the left of the runways.

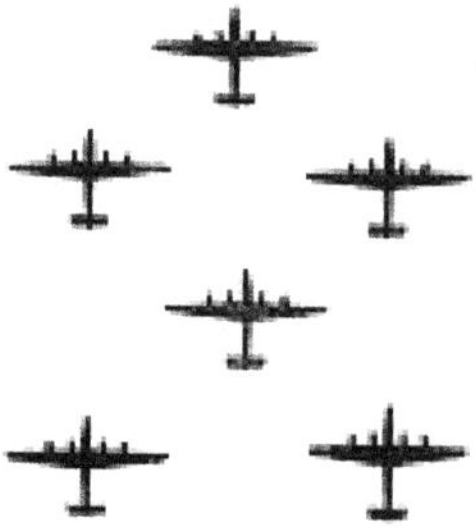

{2}

MEN

ESCAPE, EVASION & REPATRIATION

FROM THE VERY BEGINNINGS of combat missions, the Fifteenth Air Force has been operating an ever developing program to return airmen shot down behind enemy lines, in neutral territory, or downed at sea, to Allied control. Throughout its history of combat operations, personnel missing in action (except those known to have been killed) totaled 17,918. The number of men recovered from eleven countries controlled by the Axis forces, from Switzerland, and from sea rescue was 5,559.

Austria	*878*
Italy	*402*
Germany	*514*
Yugoslavia	*1,151*
Rumania	*1,434*
Hungary	*413*
Czechoslovakia	*125*
France	*105*
Bulgaria, Greece, Albania, Switzerland, Rescued at Sea, and others	*537*
Total	***5,559***

In addition, there were over 600 other Allied military or naval personnel who were rescued and brought to Headquarters, Fifteenth Air Force for processing.

The recovery of air crews was primarily a two-part operation, the recovery of personnel lost over enemy territory, and the recovery of men from the Adriatic and the Mediterranean. The first involved techniques in escape and evasion, ending in repatriation to Fifteenth Air Force bases in Italy. The second employed units specializing in air-sea rescue. As a result of their activities and continuously evolving techniques, the Fifteenth Air Force recovered more downed airmen in more countries than any other air force.

* * * * *

REPATRIATION

How airmen were recovered from enemy territory was an operation conducted by the Fifteenth Air Force and involving a number of Allied organizations in the Mediterranean. A report, known as a Missing Air Crew Report, was required to be filed, through channels, with the Escape Intelligence Officer, Fifteenth Air Force, by 0800 the day following the mission on which the aircraft was failed to return. The "MACR" was completed by Squadron S-2 for the signature of Group S-2. It included the date, squadron, type of aircraft, serial number, point of origin, target, and time and nature of trouble. It also included a full roster of all men on the aircraft, including name, rank, position, serial number. Witness statements included a brief description of events leading to the loss of the ship and its condition when last seen. Also included was the location where last seen and the number of parachutes seen. According to the given location provided on the MACR, a signal was sent by Fifteenth Air Force S-2 to and agencies or helpers in the area alerting them to look for the lost airmen, and to help conceal them from enemy forces. These agencies included the Office of Strategic Services (OSS), Balkan Air Force, and British units I.S.9, No. 1 Special Force, 37 Military Mission, Force 266, and Force 399.

All returning air crews were interrogated by the Escape and Evasion sub-section of Intelligence (A-2). This information was transformed into bulletins to train combat crews on the latest methods to evade capture, or to escape if captured. They also supervised escape and evasion training given by A-2 and S-2 officers. In addition to A-2 activities, there were a number of mass evacuations from various countries, which were conducted by the Operations Section (A-3) of Fifteenth Air Force Headquarters.

The agency that would become most involved with repatriation was the Air Crew Recovery Unit (ACRU), stationed at Bari, responsible to HQ, MAAF for operations, and to the Fifteenth Air Force for issues of administration. Prior to November 1943 (XII Bomber Command) there were 55 returnees. By the end of the year 108 men were processed by the Fifteenth Air Force. During January – March 1944 only 44 crewmen were recovered, but in April and May 1944 262 escapees and evaders were processed. During June there were 220, and during July 218 were interrogated and processed. The increase in returnees was due in large part to the corresponding increase in offensive operations. In August and September Rumania and Bulgaria capitulated, resulting in a large number of airmen brought home by well planned and executed plans named Reunion in Rumania, and Freedom in Bulgaria. As a result, during July and August, there were 2,182 men from those two countries.

ACRU, formed in July 1944, would be responsible for gathering up downed airmen, providing for medical care, and getting them assembled and ready for their return to Italy. All of this would be coordinated with Balkan Air Force Special Operations and the military missions that had been assigned to those areas

In addition to evacuating airmen, ACRU flew 61 supply sorties (49 effective), dropping 117,600 pounds of supplies, including 4,500 pounds delivered by ACRU planes landing at pre-arranged landing zones. The ACRU "midget air force" consisted of two B-25s and two L-5s plus the use of three C-47s from the 60th Troop Carrier Group.

* * * * *

CREW INTERROGATION

As mentioned earlier, when a man was repatriated to Italy his first duty was to report to the Escape Intelligence Officer. Most times the arrival had already been anticipated and the name checked against MIA records obtained from reports forwarded by the missions responsible for their repatriation. To ensure foreign agents didn't gain access to Allied facilities, a careful security process was observed. During the interrogation, for example, careful attention was paid to discussions about the airman's fellow crew members to verify he was the man he claimed to be.

When a man arrived at the 26th General Hospital or 22nd Replacement Depot, he would have been required to complete a questionnaire, including name, rank, serial number, organization MIA date, the area where he landed, and brief personal details. Also included was a thorough disinfectant, bath, and change of clothing. The next stop was the Escape Intelligence office to complete additional forms prior to the actual interrogation. Questions during the interrogation were divided into six parts:

- Personal details, aircraft difficulties, reason for bailout, and present status and location of aircraft. Whether subject was ever in enemy hands or in neutral country. If captured, name and location of prison camp. If in neutral territory, where interred. Complete story of evasion, or escape and evasion (from the bailout to return to Italy.
- Names, dates, locations of helpers, plus name and location of missions as well as evacuation points.
- A report on the effectiveness of escape aids; use, if any, of escape kits. Comments, criticisms, and suggestions, such as GI shoes, sidearms, extra woolen socks, etc.
- Names, dates, and locations of pro-Axis personalities. Any reports of atrocities committed, witnessed, or discussed. All military information such as bomb damage, enemy movements, ammunition dumps, oil storage, possible emergency landing grounds in enemy territory.
- All information concerning P/W camps, if any. Enemy commanding officer of the camp; Allied C/O. Outstanding personalities at the camp. Escape committee and attempts to escape. Treatment received, medical care, hygienic conditions, cases of brutality, food rations, hospitalization, etc. Red Cross aid boxes. Enemy interrogation methods, subterfuges, threats, tricks, etc.
- Casualty information on crew members, other Allied personnel met and where last seen. Location of graves or bodies.

A great deal of current intelligence information could be extracted and forwarded to A-2 Analysis, ACRU, I.S. 9, Operational Engineering, A-4 Movements, etc. This, in turn, was used to update information passed on to air crews as an aid in evasion, or escape if captured and highlighted. One change, resulting from comments by returning airmen, was the creation of an identification card. There had been several occasions when downed flyers encountered people who didn't necessarily believe they were actually Americans. This was especially true when it came to Russian soldiers who sometimes mistook them for Germans, defused only by the arrival of a Russian officer. This card, folded in three ways, included one part with an American Flag in full color, the airman's photo and signature, and the United States Army seal. Next appeared the English statement "I am an American airman. Please take me to your commanding officer and notify nearest American or British military mission in Belgrade, Bucharest, Poltava, or other nearby place. Also, please arrange for transportation. Thank you." Beneath this statement appeared "Show this to the Russians" along with the Russian translation of the above statement. On the reverse side in Polish, Czech, Yugoslav, Hungarian, and Austro-German is the statement "I am an American. Please take me to the nearest American or British mission, or to the nearest Russian military authority." Crews carrying these cards received an immediate benefit.

Results of these interrogations also led to changes in the escape kit. Because of the difficulty of enclosing money for all the countries within the range of Fifteenth Air Force aircraft, a "chit" system was developed where the airmen would write on a piece of paper the assistance he was given that the helper could take to an American mission for payment.

* * * * *

BIG FENCE

When in doubt, call Big Fence. That's the advice for crews who were lost or about to bail out or ditch in the Adriatic. That was the experience of Lt. Bertrand J. Arents of the 461st Bomb Group, 767th Bomb Squadron, who was on a Lone Wolf mission December 2 & 3 1944. "We were supposed to bomb at 21,000 feet. We broke out of the clouds at 18,000 feet and there was no weather above us. We didn't like it one bit being in the clear but proceeded on to the target. Our radar operator got good ground readings and in about 5 minutes we hit our I.P. Our crew saw the bomb flashings through the clouds as we hit the target. They looked like lightning strikes. I turned right in a descending turn and at that time while turning right we saw bullet strikes on our No. 1 engine. German fighters had found us and if I had not turned sharply they would have hit us in the fuselage. They shot out the engine and set it on fire. Also unfortunately they destroyed the feathering device and the propeller rotated in the opposite direction creating tremendous drag on the left side of the aircraft. This forced us to almost cut all the power to our No. 4 engine or we would be in a spiral to the left. Also we had to reduce power to our No. 3 engine. My co-pilot was able to put out the fire in the No. 1 engine with the engine fire extinguisher. We dove into the clouds and were thus free from the fighters. With only two engines operating we began to lose altitude and was concerned about the mountains in the Alps. I told the crew to throw out all the stuff they could find overboard including the waist guns to lighten the ship.

Fortunately, we had Big Fence, the radar facility at Bari, Italy. We contacted them and we were lucky as they picked us up on their radar and directed us around the highest peaks of the Alps. We constantly lost altitude and the co-pilot and I had to use both feet on the rudder control to keep the ship flying straight because of the tremendous drag of No. 1 engine. The front line of the German and British 8th Army was about 20 miles north of Rimini, Italy. We knew if we could reach Rimini we could put the ship down in a field, as we couldn't make it back to any airfield and we wouldn't be captured. We did not want to bail out either. As we neared Rimini on the east coast of Italy we were at 1,000 feet but in the clouds. We were happy in the clouds as the Germans in the front lines couldn't see us or shoot us down. Pretty soon Big Fence told us there was a big field west of Rimini that they believed we could land on. We broke out of the clouds at 500 feet and saw Rimini. We turned right and saw the field."

The 6649th Navigational Aids Squadron (Provisional) was activated on September 1, 1943, but its origin began the previous January when Lt. Donald P. White and nine enlisted men were transferred from the 561st Signal Battalion at La Senia, Algeria to Viskea, Algeria to set up and operate a VHF (Very High Frequency) Homer Station for the 1st Fighter

Group. Once the station was set up the 1st Fighter Group received their own station requiring Lt. White to dismantle what they had just built. At about this time XII Bomber Command had acquired three VHF teams and equipment to set up a VHF Triangulation System and Lt. White was given the assignment of making the new system a reality. He was able to acquire the necessary equipment, tents, transportation, etc. and began identifying sites for the VHF sets. Despite all the issues that arise creating a new program, the program was up and running by March 1943.

By January 1944, the move to Italy had been completed. Big Fence control was now in a centuries-old fortress, the Castel del Monte just north of Foggia, while the VHF sets had been located at:

- Big Fence Control, Castle del Monte, Italy
- Jacob D/F Station, Monte San Angelo, Italy
- Dark Grey D/F Station, San Marco, Italy
- Cutlet D/F Station, Pomarico, Italy
- Radio Range, Foggia, Italy

January 24, 1944 was the biggest day in Big Fence history when navigational aid was given to 70 different aircraft or flights of aircraft, and plots on nine aircraft who had ditched in the Adriatic were passed to Air-Sea Rescue. The crews of all nine aircraft were picked up by rescue boats dispatched by Big Fence. The day's accomplishments would result in a commendation from Gen. Twining. Total activity for the month was 392 homings, 23 fixes and 17 Maydays. There are no records that tell us how many 461st crews benefited from the services of the 6649th Navigational Aids Squadron, or the Air/Sea Rescue personnel, but it is safe to say a lot of good men went home because of them.

* * * * *

Air/ Sea Rescue

Successful rescues were not a matter of chance. They resulted from the careful coordination of many units as well as the training and performance of the aircraft crews.

The responsibility for rescuing our crews was vested in the Mediterranean Allied Coastal Air Force (MACAF). They provided search and rescue aircraft aided by high speed launches under the control of the Royal Navy. If needed, the Fifteenth Air Force would also provide aircraft search and cover. However, control and coordination was at all times under the authority of the Air/Sea Rescue Service of MACAF.

The Mediterranean Allied Coastal Air Force controlled all air/sea operations in the Mediterranean Theater, and had the following responsibilities:

- Fighter defense of shipping.
- Fighter defense of ports and vulnerable areas.
- Anti-submarine protection of shipping.
- Shipping reconnaissance.
- Offensive action against enemy shipping and harbor installations.
- Night intruder operations.

Off the coast of Italy this rescue boat could be searching for a downed aircraft or a capsized sailboat.

Due to the highly efficient radar, radio, and communications networks fighters, search, and rescue aircraft, strategically, were instantly available in each sector. These units operated in defined units along the coastline and were responsible for all incidents in their sector. Fighters and search & rescue type aircraft were instantly available in each sector. Whenever an air/sea rescue action arises, all available information is sent to the closest Fighter Sector Control Room. The Control Room dispatchers are required to dispatch fighters to the search area, amphibious aircraft, if available, and instruct the local naval authority to send a high speed launch (HSL) or other surface craft. The Control Room would have also assigned air/sea search and rescue aircraft to patrol the areas along the return routes of the combat formations, and assigned marine rescue ships to Prearranged Rendezvous Points (RVP's).

The pilot and Radioman also play an important role in increasing the crew chances for survival. The earlier he provides essential information to Air/Sea Rescue Operations, the greater the chances of being picked up quickly. With enough warning some crews were actually spotted before they hit the water. There were two types of messages required of the plane in jeopardy – "Urgency" and "Distress. An Urgency message is made at the earliest sign of problems. The pilot believes he may have to make an emergency ditching or landing, (not immediate). Early contact allows the D/F stations to get a good fix on the aircraft and alert rescue planes and ships in case ditching is required. A Distress message is used when ditching is expected within 10 minutes.

It was learned, both by experience and experiment, that it is important to jettison all heavy equipment, and for most aircraft, approach the landing at a three-point altitude with the tail slightly down. However, the B-24 should be ditched level without putting the tail down. In either case, it was important to land into the wind. Ditching results in far more rapid deceleration than a crash landing on land so minimal speed at the moment of impact was absolutely important. All hatches had to be sealed except those that would be used for escape. Crewmembers had to be braced for the impact, and since there would be two impacts (one when the tail hits the water and a much stronger second impact when the fuselage hits the water, had to remain braced until after the second impact. Bracing had to involve either of two positions; either in a sitting position with the feet pressed against a solid object and the back and head against a bulkhead high enough to protect the head, or laying down with the head toward the tail and feet firmly braced against a bulkhead.

The receiving stations plot the position of the aircraft and obtain as much information as possible, sending it to the nearest Fighter Sector Control Room. The Fighter Controller, in turn, passes the information to air/sea rescue and navy authorities, including the object of the search, area to be searched, the type, call sign, and frequency, and the position and movement of all cooperating ships and aircraft.

Two sailors take a break on their Air/ Sea Rescue boat.

Upon sighting survivors in the water, the air/sea rescue aircraft would circle the men in the water, while transmitting details to their command. They may land and pick up the downed crew, continue to circle until surface vessels arrive, or vector a high-speed launch to their position. Should surface vessels pick up the survivors, they would transmit full details to their naval command and advise the search aircraft of their actions. They also transmit their arrival ETA, which would be forwarded to concerned authorities.

When the search is unsuccessful, the Fighter Controller can initiate an international distress call on 500 kc, providing full details to all possible rescuers, Allied, Neutral, or Enemy in the hope of increasing the odds of survival for the crew.

However, should the aircraft manage to reach land, the Fighter Controller would notify the closest Flying Control Office and inform them of actions taken. In this case Big Fence would again attempt to find a landing place for the crew.

KRIEGSGEFANGENEN (Prisoners of War)

GERMANY OPERATED PRISONER-OF-WAR CAMPS (Kriegsgefangenenlager) throughout Germany and occupied countries. Many of these camps were designed for specific prisoners such as civilians (Intemierungslager or Internment camp), camps for Naval personnel (Marine-Lager), or Merchant Marine camps (Marine Intemierten Lager). There were also Officers Camps or Offizier Lager, and camps for Enlisted Personnel or Stalags. Allied airmen hoped to be sent to camps under the control of the Luftwaffe (German Air Force) as they generally received better treatment than they could expect in other camps. The Stalag-Luft (Luftwaffe-Stammlager) was administered by the Luftwaffe for Allied aircrews.

Dulag Luft was a transit camp known officially as Durchgangslager der Luftwaffe (Transit Camp of the Luftwaffe) which contained an interrogation center and hospital. Located in Oberursel, it was considered to be the greatest interrogation center in Europe. The vast majority of American and Allied airmen were sent here for interrogation before being transferred to a permanent prison camp. Prisoners were kept in solitary cells and studied by interrogators to determine then best methods to extract information. Many received very harsh treatment including threats of torture, being left for long hours in solitary confinement, provided with little food, or had medical care withheld. The level of treatment depended on individual interrogators, some of which used brutal methods. However, a few were humane. Hanns Scharff, for example, always treated prisoners with respect, yet was able to extract answers from them through conversation without the prisoners realizing they were providing valuable information. There were also many documented cases of Scharff providing care for sick prisoners and protecting others from the SS. While he was only a Corporal, Scharff always wore civilian clothing, causing prisoners to believe he was a high-ranking officer. With testimony from his former prisoners Scharff was later allowed to move to the United States and counted Gen. Jimmy Doolittle among his friends. Unfortunately, Scharff was the exception among interrogators.

Later in the war Dulag Luft was moved closer to Frankfort and was renamed Dulag Luft West. It isn't always possible to determine which POW camp a prisoner may have been assigned due to the number of missing documents over so many years. In that case, prisoners with no known permanent POW camp have been assigned to either Dulag Luft or Dulag Luft West because they would have been assigned there at least temporarily.

After the war, the United States occupied Dulag Luft and renamed it Camp King, which began as an intelligence post. It later became a transportation post until it was closed in 1993. The camp was later torn down, replaced by a housing area.

> *"They started out by demanding all sorts of information. Who we were, where we came from, and the names of my crewmates; because they wanted to make sure they had captured all of us. They also wanted to know what targets we were bombing, what armament we carried, how many bombs, and how many bombers in our squadron. I gave them my name, rank, and serial number, and that was all*

they got out of me. They said, "Oh we have ways to make you talk," and sure enough they shut me away in a little black solitary cell, absolutely dark, just big enough for a bed. It then became psychological warfare, when they'd come in at a moment's notice, shine lights in your face and ask the same questions over and over. I kept giving the same answers, and they said, "Well that's all right, you'll just have to stay here for the rest of the war." The thought of that was disheartening to say the least. After about four days of that, they transferred me to the next camp where I was reunited with my crewmembers, and I got to see the blue sky and the sun. What a treat!

Hjalmar Johansson
461st Bombardment Group
767th Bomb Squadron

Stalag Luft I, located near Barth, Germany beginning in 1941 for Allied airmen. Between January 1944 and April 1945 the number of POWs rose from 507 to approximately 9,000 American, British, and Canadian flyers. The prisoners were ordered to evacuate the camp on 30 April 1945 with Russian troops approaching. However, the senior American officer refused to leave the camp and the Luftwaffe guards abandoned the camp, which was liberated by the Russians later that night. The Americans were evacuated to Camp Lucky Strike for shipment back to the United States.

Stalag Luft III, which was 100 miles southeast of Berlin held primarily officers. The location had been chosen because of the difficulty tunneling through the sandy soil. Nevertheless, there were two successful escapes by tunneling. One of those escapes was the basis for the movie "The Great Escape."

Stalag Luft IV, located in Gross Tychow, Poland, held enlisted men. It opened in May, 1944 with 1,500 POWs, but would eventually increase to nearly 10,000. While prisoners complained of serious problems at the camp, they were forced to tell their families that there were no problems and everything was fine. In February 1945, the prisoners were required to march on what would be known as the "Black March" to avoid advancing Russians. Lasting 86 days, the prisoners were beaten, starved, victims of exposure, and were forced to drink water from ditches. They tried to trade whatever they had with local farmers for food. Forced to eat charcoal to stop dysentery, disease ran rampant through the men.

Stalag Luft VI in Heydekrug, Germany held enlisted men. Originally opened in 1939 under another name, it was renamed in 1943 when British and Canadian men were imprisoned. Americans were added in 1944. These men were also forced to participate in the forced marches when Russian troops were approaching the camp and were treated with the same brutality as POWs from other camps. After the Russians occupied the camp, it was renamed Gulag 3 and used to hold German forces.

Stalag Luft VII-A, north of the town of Moosburg, Germany, held both officers and enlisted men, and at 86 acres, was the largest POW camp in Germany. Also, a transit camp, thousands of prisoners passed through on their way to other camps. At its liberation on 29 April 1945, there were 80,000 prisoners, most of which were Russian and French.

Stalag Luft XVII, located in Krems, Austria, was home to American enlisted personnel. The Americans occupied five compounds, each 75 X 175 yards. The barracks at Krems were built to hold 240 POWs, but at least 4,000 were crowded into the prisons within three months of its opening in 1940.

Life behind the wire depended to some degree upon which camp a prisoner was assigned. While a Stalag Luft was usually better than other camps, it was hardly a pleasant experience. Their barracks ranged from wood to concrete cells holding eight to forty men in each room. Barracks life was boring in the extreme although the Kriegies could hold sporting events, often had access to a library, produced plays, or conducted classes.

Food was always scarce, especially toward the end of the war. One of the primary tasks of the Red Cross was the delivery of food parcels to each prisoner once a week. Over 27,000,000 boxes were assembled in the United States. While most camps allowed the free distribution of boxes, some Germans collected the boxes to be doled out over the week.

The Germans were far from being gracious hosts. Discipline was strict, there was often no hot water, the barracks were cold and drafty, and the food was limited.

"Conditions were pretty rough. We were infected with fleas and lice. Without hot water, I had only one shower in five months. I didn't need a knife or fork, just a tin cup and

spoon, and that was all you needed. Whatever food came, it was ladled out into your cup, and you took whatever they gave you. In the prison camp there weren't any mice or other small animals running around – they would have been put in the soup pot immediately. What we ate or didn't eat was of primary importance. When your stomach is empty all you think about is food."

Hjalmar Johanssen
461st Bomb Group
767th Bomb Squadron

Punishment for attempted escapes could be severe even though against the Geneva Convention. Sometimes the possibility of death was present without committing any transgressions:

"One day the fliers (and only the fliers) were told to line up outside for a special roll call. None of us fliers had been prisoners very long and hadn't acquired knowledge of any of the German language – so we didn't understand what was really happening. Some of the infantrymen in the barracks had been prisoners for several years and spoke fluent German. They told us, later on, what we had been through. It seems Hitler had become incensed over the success of the Allied bombing and had ordered every airman prisoner of war to be shot! So – here we were in blind innocence, not knowing how close we were to being executed. Fortunately, the Commandant of the compound refused to obey the order. It was near the end of the war and the utility of the order, and fear of reprisal kept the Germans from executing us. I presume the word was soon received by Hitler that no one was going to carry out his order and, after standing in formation for an hour or so, we were dismissed to return to the barrack."

Trefry A. Ross
461st Bomb Group
765th Bomb Squadron

Camp Shumen the pro-communist government in Bulgaria during WWII fought against American and British forces in what the Bulgarians called the "symbolic war." During this period, a number of Allied aircraft were lost resulting in 300 airmen killed and 342 becoming POWs.

The first 10 survivors were kept in Sofia, but Allied bombing convinced the Bulgarians to move their prisoners to a safer locale location in the country.

Camp Shumen, Bulgaria

Shumen was only active for 379 days from 25 November 1843 to 8 September 1944 when it held 291 American airman, the English and Scottish flyers, and three from the South African RAF. There were also five British agents there, including one English Major, one Greek and one Serbian paratrooper, and two Dutchmen, who were also brothers.

Living conditions for the prisoners was difficult. There was no electricity or running water and medical care was sub-standard. Still, there were no deaths from malnutrition or mistreatment. The guards ate the same quantity and quality food as the prisoners. The Shumen Garrison Commandant had ordered all prisoners would be treated as members of great countries. There was also no evidence of discrimination against any prisoner, including Native American and African American POWs.

On 8 September 1944, the Bulgarian government released all their prisoners and moved them to the train station in Shumen. Early the next day they were moved to Turkey where they were flown to Egypt by the British. At that point they were under the care of the American Red Cross in Cairo until they were picked up by planes from the 484th Bomb Group who flew them to Cerignola, Italy.

* * * * *

KNOWN 461ST PRISONERS OF WAR

SQUAD.	NAME	RANK	SERIAL NUMBER	DATE REPORTED	PLACE DETAINED	CAMP	HOME STATE
766	ACE KEITH F	2Lt	0-558003	2/13/1945	AUS	Stalag Luft III	WI
764	ADELMAN MELVIN A	2Lt	0-723255	7/25/1944	GER	Stalag Luft I	IL
764	ALLEN ALFRED H	T/Sgt	38305398	5/10/1944	AUS	Stalag Luft IV	TX
	ALLEN ROY V	Sgt	18184876	5/24/1944	AUS	Stalag Luft III	TX
765	AMBROSE, ROBERT J		0-799059	5/31/1944			
766	ANDERSON GERALD W	S/Sgt	17142953	8/24/1944	ITA	Stalag Luft IV	MN
	ANDREWS FREDERICK C	S/Sgt	12024145	8/17/1944	YUG	Stalag Luft IV	NY
	ANTHONY JOHN D	Sgt	34162073	8/23/1944	GER	Stalag Luft IV	AL
	ARROTTA ALBERT T	Sgt	11120703	12/17/1944	GER	Stalag Luft I	MA
	ARROWOOD PALMER D	Sgt	35769407	12/17/1944	GER	Stalag Luft I	KY
	ASHBY DAVID A JR	2Lt	0-669156	5/24/1944	AUS	Stalag VIIA	TN
764	ASHWORTH PAUL R	F/O	T-137615	4/25/1945	AUS	German Civ. Hosp	UT
764	AUSTIN JACK J	Sgt	36867317	1/31/1945	AUS	Stalag Luft IV	MI
765	AVERY, RICHARD E		0-808933	5/31/1944			
764	BACA JOE R	S/Sgt	18220765	5/10/1944	AUS	Stalag Luft IV	NM
	BACON FLOYD A	S/Sgt	17099963	10/4/1944	ITA	Stalag Luft III	KS
	BARKS JOHN P	Sgt	11064289	12/17/1944	GER	Stalag Luft I	MA
	BARNES LLOYD E	PFC	17032175	7/18/1944	BUL	Camp Shumen	AR
	BARNHART K E JR	2Lt	0-718991	8/7/1944	GER	Stalag Luft III	CA
	BATTAGLIA RUSSELL J	S/Sgt	32735829	2/13/1945	AUS	Dulag Luft West	PA

764	BAUER HAROLD J	Sgt	33681758	10/4/1944	GER	Stalag IVB	PA
	BAUMANN EDWIN	2Lt	0-681308	5/31/1944	GER	Stalag VIIA	CA
764	BEAN, JAMES O		0-794458	4/13/1944	HUN		CA
	BEHLING GORDON C	F/O	T-004502	12/17/1944	GER	Dulag Luft	WI
764	BENTON JAMES W JR	1Lt	0-747192	7/25/1944	GER	Stalag Luft I	OR
	BENTZ FORREST M	Sgt	33486662	1/8/1945	YUG	Stalag Luft III	PA
	BESSER VICTOR H	2Lt	0-772931	8/23/1944	AUS	Stalag Luft III	CA
765	BINDRIM ARTHUR J	S/Sgt	34202986	5/31/1944	RUM	Stalag Luft III	FL
	BIRD FRANK S	1Lt	0-759343	8/26/1944	YUG	Stalag VIIA	AL
	BLACKER CHARLES L	T/Sgt	153322702	11/20/1944	GER	Dulag Luft	MI
	BLALOCK EARL D	S/Sgt	34128099	4/13/1944	HUN	Stalag Luft III	SC
	BOGGS KENNETH S	2Lt	0-690354	5/24/1944	GER	Dulag Luft	PA
	BOMBARDIER GERALD A	Cpl	11106529	5/24/1944	AUS	Dulag Luft	VT
	BONEWELL CARLOS J	Sgt	18129569	8/17/1944	YUG	Stalag Luft IV	OK
	BOWYER EDWIN W	1Lt	0-544019	7/25/1944	AUS	Stalag Luft IV	OK
765	BLALOCK, EARL D		34128099	4/13/1944			
764	BLOXOM CLARENCE W		0-668303	10/8/1944	RUM		CA
764	BRAUN WILLIAM R		0-716707	7/25/1944	RUM		NC
	BRICE RICHARD D	Cpl	38479611	8/24/1944	ITA	Stalag Luft IV	TX
765	BRINA ROBERT R	2Lt	0-722991	10/4/1944	GER	Stalag Luft I	NY
765	BRLANSKY JOSEPH C	Sgt	13157945	4/24/1944	RUM	Stalag Luft III	PA
	BRUSSO ROBERT F	S/Sgt	32383384	6/11/1944	RUM	Camp Shumen	NY

	BURKE WILLIAM	Major	0-409947	7/25/1944	AUS	Dulag Luft	NY
	BYERS THOMAS F	Cpl	33714183	12/16/1944	ITA	Dulag Luft	NY
764	CAIN ALFRED JR	Sgt	35760647	1/31/1945	GER	Stalag Luft IV	MD
	CAPALBO FREDERICK B	2Lt	0-825560	12/17/1944	GER	Stalag Luft I	CT
	CARLOCK ALBERT F	T/Sgt	39455785	6/26/1944	GER	Bratislava POW Camp	WA
765	CARON FRANK P	S/Sgt	39269401	5/24/1944	AUS	Stalag Luft IV	CA
	CAROSELLI, PHILIP J	LT	0-915791	4/13/1945	GER	Oflag Luft III	NJ
765	CARF, RUDOLPH C		0-1095418	7/22/1944			
764	CARROLL EDGAR R	Sgt	12239327	1/31/1945	AUS	Dulag Luft West	NY
	CARTER, RICHARD	2Lt	0-555652	8/22/1944	YUG	Stalag Luft III	UT
	CASH HUGH I	Sgt	14153842	12/17/1944	GER	Stalag Luft I	NC
764	CATALANO ANTHONY	T/Sgt	12191903	5/10/1944	GER	Stalag Luft IV	NJ
	CATANA ANTHONY M	2Lt	0-695801	7/25/1944	AUS	Stalag Luft I	NJ
764	CHABIN HARRY	S/Sgt	13046578	5/10/1944	AUS	Stalag Luft IV	PA
764	CHAKLOS JOHN F		0-154374	1/19/1945	YUG		MI
	CHAN EDWARD N	2Lt	0-2060774	12/17/1944	GER	Stalag Luft I	NY
764	CHARLAND DONALD G	Sgt	36456233	7/25/1944	AUS	Stalag Luft III	MI
	CHESHIER ELTON E	S/Sgt	18110627	7/25/1944	AUS	Stalag Luft IV	TX
	CHILDS CLAYTON A JR	T/Sgt	12199335	4/13/1944	AUS	Stalag VIIA	NY
764	CHILDS HOMER D	Sgt	11101348	4/2/1944	GER	Stalag 17B	CT
	CHOJNOWSKI EDWARD J	2Lt	0-2058706	1/31/1945	AUS	Stalag Luft West	IL

	CLARK BEVINS	S/Sgt	15108354	6/26/1944	AUS	Stalag Luft IV	IN
765	CLAY CHARLES R	S/Sgt	19165159	12/17/1944	GER	Stalag Luft I	CA
	CLIPPERT EMORY C	Sgt	36888974	10/4/1944	GER	Stalag 7A	MI
	COCKING BYRON D	2Lt	0-810828	6/11/1944	RUM	Camp Chumen	PA
765	CODY JACK R	1Lt	0-560800	12/17/1944	GER	Stalag Luft III	NM
765	COONAN ROBERT A	Sgt	20347800	7/25/1944	AUS	Stalag Luft IV	MD
765	CROSSMAN PHILLIP J	2Lt	0-709511	12/17/1944	GER	Stalag Luft I	OH
765	CULLER BUFORT E	S/Sgt	14181033	5/31/1944	RUM	Stalag Luft IV	NC
	CUMMING ROBERT D	S/Sgt	38103221	7/12/1944	FRA	Stalag Luft IV	CO
765	CURTIN JOSEPH S	S/Sgt	32723135	5/31/1944	YUGO	Stalag Luft IV	NY
764	CUSSEN EDWARD E	S/Sgt	39085129	5/10/1944	AUS	Stalag IXB	IA
764	DALLAS, CHRIST	2Lt	0-2002032	3/26/1945	AUS	Stalag Luft VIIA	MI
764	DAMANTE ARMANTE R	1Lt	0-678411	7/25/1955	GER	Stalag Luft I	NH
	DAVIS FREDERICK E	Sgt	37501592	8/22/1944	GER	Stalag Luft IV	KS
764	DAVIS, RICHARD C.	1LT	0-685555	3/23/1945	Aus	Moosberg N. IX	VT
	DAVIS ROBERT E	Sgt	34700197	10/4/1944	ITA	Stalag Luft III	AL
	DEAN HARRY E	Sgt	15377793	4/13/1944	HUN	Stalag VIIA	WV
	DEMARCE DONALD H	2Lt	0-768478	12/18/1944	GER	Stalag Luft I	CA
765	DERUDDER, PETER J		0-735286	5/31/1944			
765	DESPAIN GLEN P	T/Sgt	36024888	5/31/1944	RUM	Stalag Luft III	IA
765	DIAMOND GEORGE	S/Sgt	12187570	7/25/1944	AUS	Dulag Luft	NY
	DIGGS WILLIAM E	2Lt	0-756794	5/24/1944	AUS	Stalag Luft III	IL

	DILLON IVO L	T/Sgt	34702710	7/25/1944	AUS	Stalag Luft IV	TN
	DIXON LAWRENCE F JR	T/Sgt	34607342	12/17/1944	GER	Stalag Luft III	NC
	DORNEY DIC L	2Lt	0-712779	9/14/1944	YUG	Stalag Luft III	OH
	DRAPER HAL D JR	2Lt	0-776885	12/17/1944	GER	Stalag Luft I	CA
	DRENA MICHAEL	S/Sgt	31049732	2/13/1945	AUS	Stalag Luft West	CT
	DUNHAM HARRY W	T/Sgt	6980505	12/16/1944	ITA	Stalag Luft I	MA
	DUPREE ARTHUR R	S/Sgt	31226428	12/18/1944	GER	Stalag Luft I	MA
	ECHTERNACHT NORRIS G	2Lt	0-713371	8/17/1944	RUM	Stalag Luft III	CA
764	EDMISTON HARRY D	2Lt	0-2061225	12/17/1944	GER	Stalag Luft IV	CO
765	EIDSMORE LAWRENCE M	Sgt	32312554	12/17/1944	GER	Stalag Luft XII	NY
	ELDER THOMAS L JR	2Lt	0-694909	7/25/1944	AUS	Stalag Luft I	AL
765	ENGBROCK GLENN H	2Lt	0-723331	7/25/1944	AUS	Stalag Luft I	TX
	ENGELKE WILFRED A	Sgt	38414209	12/17/1944	GER	Stalag IIIA	TX
765	ESSER GEORGE	S/Sgt	39557583	7/25/1944	AUS	Stalag Luft IV	CA
	ESTERLINE WILLIAM B	S/Sgt	35542984	8/25/1944	FRA	Dulag Luft	OH
765	FARR HOWARD J	S/Sgt	16138734	7/25/1944	GER	Stalag Luft IV	IL
765	FARRELL JEFFSON L JR	S/Sgt	14161846	5/31/1944	RUM	Stalag Luft III	TN
	FEDRAU ABRAHAM	S/Sgt	39404271	7/25/1944	AUS	Stalag Luft IV	CA
764	FEENEY JAMES B	Pvt	14025848	10/4/1944	GER	Stalag Luft IV	MA
764	FEILER LAURENCE G JR	Cpl	36884529	7/25/1944	GER	Stalag Luft III	ID
	FERGUSON BOBBIE C	Sgt	17169724	7/25/1944	GER	Stalag Luft IV	IA
	FINE DALE V	Sgt	18163238	4/2/1944	GER	Stalag 17B	OK

765	FISHER MANLY K	Sgt	12209083	7/25/1944	AUS	Stalag Luft IV	NY
	FISHER ROBERT H	2Lt	0-812532	7/25/1944	AUS	Stalag Luft I	ID
	FORSTER JOHN G	S/SGT	11081714	3/23/1945	GER	Moosberg VIIA	MA
764	FOSS CHARLES F	Cpl	31360952	12/17/1944	GER	Dulag Luft West	ME
	FOX HOWARD R	2Lt	0-723044	8/17/1944	YUGO	Stalag Luft III	PA
	FRANCISCO, V.L. SR.	2LT	0-2059020	12/17/1944	GER	Stalag Luft I	AR
764	FRANKLIN ANDREW D	S/Sgt	14094810	7/25/1944	AUS	Dulag Luft	AL
764	FREEMAN RICHARD E	1Lt	0-812566	7/25/1944	GER	Stalag Luft I	PA
765	FULTON ORAN E	F/O	T-003309	7/25/1944	AUS	Stalag Luft I	CA
764	FUNK ANDERSON A	T/Sgt	36653208	10/4/1944	GER	Stalag VIIA	MI
	GALVAN ROBERT A	1Lt	0-672598	12/17/1944	HUN	Stalag Luft I	TX
	GAVORA JOHN JR	2Lt	0-691983	6/11/1944	RUM	Camp Chumen	NJ
765	GEIER LEONARD G	Sgt	12156010	12/17/1944	GER	Stalag Luft III	NY
	GEORGE EDWARD K	1Lt	0-705929	12/18/1944	GER	Dulag Luft West	FL
	GIBBONS JOHN F	S/Sgt	12035354	4/16/1944	RUM	Stalag 7A	NY
764	GIBSON WILLIE	S/Sgt	15339626	7/25/1944	GER	Stalag Luft IV	IN
765	GIDEZ MELVIN D	2Lt	0-825834	7/25/1944	AUS	Stalag Luft I	MA
	GILDART JOHN R	S/Sgt	11053056	4/16/1944	YUG	Stalag VIIA	ME
	GILLIS WELLINGTON A	2Lt	0-831151	1/8/1945	YUG	Stalag IIIA	MA
	GILMER CLARENCE L	Cpl	19201148	1/8/1945	YUG	Stalag Luft XII	OR
	GITHENS KENNETH O	2Lt	0-759219	7/25/1944	AUS	Stalag Luft I	OR
765	GLASSON JAMES A JR	Cpl	38554034	1/8/1945	YUG	Dulag Luft West	TX

765	GODINO PETER N	S/Sgt	7021152	7/25/1944	GER	Stalag Luft IV	PA
	GOLDSTEIN GERALD	S/Sgt	12190044	4/2/1944	YUG	Stalag Luft III	NY
	GOULD DWAINE E	2Lt	0-716861	8/24/1944	ITA	Stalag VIIA	AZ
	GRANCHELLI FELIX	S/Sgt	11099567	8/7/1944	GER	Stalag Luft IV	MA
764	GRANGER URBAIN H	S/Sgt	36397799	12/17/1944	GER	Stalag Luft I	IL
	GRAYSON CECIL J	Cpl	34770593	8/22/1944	AUS	Stalag Luft IV	NC
764	GREBE LEROY H	S/Sgt	38163454	3/9/1945	YUG	Dulag Luft West	TX
	GRZYWINSKI ZYGMUND H	Pvt	36506060	7/25/1944	AUS	Stalag Luft IV	MI
765	GUY R M	2Lt	0-717782	8/3/1944	ITA	Stalag Luft I	TX
	HAGGERTY PAUL T	Cpl	31310455	12/16/1944	IT	Stalag Luft I	MA
	HAMILTON JEFF L JR	1Lt	0-673556	12/17/1944	GER	Stalag Luft I	TX
765	HANLEY HUGH H	F/O	T-003731	12/17/1944	GER	Stalag Luft I	NJ
765	HARP LELAND J	2Lt	0-703876	7/25/1944	AUS	Stalag 398	CA
764	HARRISON DONALD B	T/Sgt	37499822	10/4/1944	GER	Stalag VIIA	MO
765	HARRISON KERMIT C	S/Sgt	37085916	7/25/1944	AUS	Stalag Luft IV	WY
	HASEK FRANCIS J	2Lt	0-719066	8/24/1944	ITA	Stalag VIIA	NJ
	HAWKINS MALCOLM C	2Lt	0-2057296	12/17/1944	GER	Stalag Luft I	TX
765	HAWTHORNE FRANK G	Sgt	15171553	10/4/1944	GER	Stalag Luft IV	OH
	HAYNES HARRY L	Cpl	18242887	8/23/1944	NOR	Stalag Luft IV	TX
764	HEALD ROBERT L	1Lt	0-799399	6/11/1944	RUM	Camp Chumen	NJ
	HELTON ELDRED H	2Lt	0-826433	12/17/1944	GER	Stalag Luft XII	NC
	HENNON ERVIN R	Sgt	35090007	2/13/1944	AUS	Dulag Luft West	FL

	HEROLD ROBERT A	1Lt	0-708036	1/19/1945	YUG	Stalag Luft III	OH
	HERSHKOWITZ B H	F/O	T-132286	1/8/1945	YUG	Stalag IIIA	CA
	HEWITT, ARTHUR L	2Lt	0-1691610	7/25/1944	AUS	Stalag Luft III	TN
	HILL WILLIAM B	T/Sgt	34606592	7/25/1944	AUS	Stalag Luft IV	NC
	HOELTGEN FRANCIS B	1Lt	0-455635	7/12/1944	FRA	Stalag Luft I	MN
764	HOGAN ROBERT E	S/Sgt	15105886	5/10/1944	AUS	Stalag Luft IV	IN
764	HOKR FRANK V	2Lt	0-2064256	12/17/1944	GER	Stalag Luft XII	IL
764	HOLLERN JOSEPH D	Sgt	13158659	1/31/1945	AUS	Dulag Luft West	PA
	HOLLINGSHEAD H K JR	2Lt	0-772117	7/22/1944	RUM	Dulag Luft West	IA
	HORSTMANN DONALD G	2Lt	0-709491	8/24/1944	ITA	Stalag Luft III	IA
	HOUCK ROY V	Cpl	35333253	8/23/1944	AUS	Stalag Luft IV	OH
765	HOUSEHOLDER JOHN H	Sgt	35380587	4/24/1944	RUM	Stalag Luft III	OH
	HOVEY ROBERT B	2Lt	0-1286838	4/13/1944	HUN	Stalag Luft III	IL
764	HUMPHRESS CHARLES A	Sgt	35374314	4/13/1944	HUN	Stalag Luft III	IN
	HYLTON DONALD W	2Lt	0-755351	5/24/1944	AUS	Dulag Luft	CA
764	HYMBAUGH HOMER E	Cpl	16073922	12/17/1944	GER	Stalag Luft I	IL
765	JACKSON WILLIAM M	2Lt	0-715090	10/4/1944	GER	Stalag Luft I	DC
	JACOBSEN LYMAN	Sgt	11079168	10/4/1944	GER	Stalag Luft IV	ME
765	JAMES DONALD R	S/Sgt	13002659	7/25/1944	AUS	Stalag Luft IV	PA
	JAROS ARTHUR C JR	2Lt	0-718324	8/23/1944	GER	Dulag Luft	NY
	JAZWA WALTER E	S/Sgt	32518264	5/24/1944	AUS	Stalag Luft IV	NY
	JOHANSSON HJALMAR O	Cpl	32984706	12/17/1944	GER	Dulag Luft	NY

	JONES ALBERT G	Sgt	19203495	12/17/1944	GER	Stalag Luft III	CA
	JONES CAPTAIN E JR	2Lt	0-755356	7/25/1944	AUS	Stalag Luft I	TX
	JONES, ROBERT	F/O	T-123938	7/25/1944	GER	Dulag Luft	WA
	JONES THOMAS E JR	Sgt	14065075	8/24/1944	AUS	Stalag Luft IV	SC
764	JORDAN JAMES M	S/Sgt	15338967	7/22/1944	RUM	Dulag Luft	OH
	JUNGE PAUL F	Sgt	6480126	7/22/1944	RUM	Dulag Luft	IL
765	KAMPS EVERETT C	2Lt	0-696102	5/31/1944	RUM	Stalag VIIA	NY
	KANE JOHN J	2Lt	0-681434	7/25/1944	AUS	Dulag Luft	IL
	KARALIS GUST A	Sgt	36648106	5/24/1944	AUS	Dulag Luft	IL
765	KASOLD EDWARD A JR	2Lt	0-714116	12/17/1944	GER	Dulag Luft West	CA
	KAVANAUGH HOWARD	Sgt	32862925	8/7/1944	GER	Stalag Luft IV	NY
	KEENEY THOMAS H	2Lt	0-713469	10/16/1944	ITA	Stalag Luft III	MO
	KEFFER WARREN G	S/Sgt	39333985	8/17/1944	RUM	Stalag Luft IV	OR
764	KENDLE KENNETH K	S/Sgt	15377323	7/25/1944	AUS	Stalag Luft IV	OH
764	KHUN WILLIAM W	SGT	16097864	3/9/1945	AUS	Stalag Luft IV	NJ
	KILGORE JOHN C	S/Sgt	16046543	12/18/1944	GER	Stalag Luft IV	WI
	KIMBALL WALLACE C	S/Sgt	39011117	4/16/1944	GER	Stalag VIIA	CA
765	KLARSFELD MILTON A	F/O	T-127713	12/17/1944	GER	Stalag Luft I	NY
765	KLEPPER EDWARD J	Sgt	32889999	10/4/1944	GER	Stalag Luft IV	NY
764	KNAUBER KENNETH	Cpl	39691590	7/25/1944	GER	Dulag Luft	CA
	KOEPPEL HAROLD G	2Lt	0-770689	7/25/1944	AUS	Dulag Luft	MN
	KOESTER JOHN	Sgt	37615715	7/25/1944	AUS	Stalag Luft IV	FL

765	KOESTER ROBERT R	2Lt	0-716470	8/3/1944	ITA	Stalag Luft III	IL
	KOISTINEN PAAVO A	2Lt	0-755362	7/25/1944	GER	Stalag Luft I	NJ
	KOONTZ ROGER A	Cpl	13189919	12/17/1944	GER	Stalag Luft I	PA
764	KOSIEROWSKI S A	Sgt	32810179	4/13/1944	HUN	Stalag VIIA	NY
	KOSTEN ALEXEY J	S/Sgt	19187766	12/18/1944	GER	Stalag Luft IV	CA
765	KREPS EARL W	2Lt	0-822200	12/17/1944	GER	Stalag Luft XII	PA
	KUDLA TONY J	S/Sgt	37274648	7/25/1944	GER	Stalag Luft IV	MN
	KUHAR MIKE G	S/Sgt	35318882	5/24/1944	AUS	Stalag Luft IV	OH
	KUHLMAN ROY F	2Lt	0-713477	10/16/1944	ITA	Stalag Luft III	MN
	KUSSLER EDWARD A	2Lt	0-2057763	12/17/1944	GER	Dulag Luft	IL
	LACLEF DENSAL E	S/Sgt	37208899	6/11/1944	RUM	Camp Chumen	KS
	LAMB WILLIAM G JR	S/Sgt	15119640	8/7/1944	GER	Stalag Luft IV	KY
	LAULIS JAMES C	2Lt	0-692029	6/26/1944	AUS	Dulag Luft	WV
764	LEASURE GEORGE A	2Lt	0-2064278	3/26/1945	AUS	Dulag Luft	KS
764	LEATHERMAN D T JR	Sgt	33719965	10/4/1944	GER	Dulag Luft West	MD
	LEETH GEORGE D	S/Sgt	38038423	12/18/1944	GER	Dulag Luft West	TX
	LETENDRE PAUL L	S/Sgt	31266273	7/25/1944	AUS	Dulag Luft	NH
764	LEVINE ABRAHAM	1Lt	0-801110	7/25/1944	AUS	Stalag Luft I	NY
	LINHARES CARL J	S/Sgt	19002136	6/11/1944	RUM	Camp Chumen	CA
	LIZOTTE ROBERT A	S/Sgt	11085542	6/11/1944	RUM	Camp Chumen	VT
	LOGUE WILLIAM H	1Lt	0-671458	7/25/1944	AUS	Dulag Luft	TN
764	LORING JOSHUA JR	2Lt	0-718113	10/4/1944	GER	Stalag Luft I	MA

	LOVE JAMES	2Lt	0-684176	4/16/1944	YUG	Stalag Luft III	NY
	LOWE SAMUEL F JR	1Lt	0-805308	7/25/1944	AUS	Dulag Luft	GA
765	LUDLUM WARREN O	2Lt	0-813535	7/25/1944	AUS	Dulag Luft	NJ
765	LYKE RICHARD G	S/Sgt	39193545	4/13/1944	RUM	Stalag Luft III	WA
	LYONS THOMAS S	Cpl	14153678	12/17/1944	GER	Stalag Luft III	TN
764	MAC DIARMID ROBERT A	1Lt	0-722336	1/19/1945	YUG	Stalag Luft III	NV
	MAC DONALD THOMAS H	Cpl	31221201	12/16/1944	ITA	Stalag Luft I	ME
765	MAHNKE ARTHUR	S/Sgt	36707778	3/9/1945	AUS	Dulag Luft	IL
	MAFUTA EUGENE A	S/Sgt	16156435	7/15/1944	RUM	Dulag Luft	WI
764	MANDLY FREDERICK A	Sgt	31009218	10/4/1944	GER	Stalag Luft IV	CT
	MANN FRANKLIN R	Sgt	34762479	2/13/1945	AUS	Stalag Luft III	GA
764	MANNION REGIS P	Sgt	13170881	10/4/1944	GER	Stalag Luft IV	PA
765	MARTIN CHARLES M JR	S/Sgt	35449021	4/24/1944	AUS	Stalag Luft IV	WV
	MARTIN DONALD M	T/Sgt	16027324	1/20/1945	AUS	Dulag Luft West	IL
	MARTIN WINFIELD D	Cpl	11098596	8/23/1944	GER	Dulag Luft Hosp.	VT
	MARVILLE T T 3RD	Sgt	32753011	8/22/1944	GER	Stalag Luft IV	NJ
	MC COY CLIFFORD A	Sgt	37393128	4/2/1944	YUG	Stalag 17B	MO
	MC CULLEY NEIL W	2Lt	0-719112	12/18/1944	GER	Dulag Luft West	FL
764	MC CULLOCH LEROY S	Cpl	18199621	7/25/1944	GER	Dulag Luft	TX
	MC CULLOUGH HOWARD E	Sgt	36449902	8/23/1944	GER	Stalag Luft lag XII	IL
	MC DANIEL DONALD S	Sgt	35769385	11/20/1944	AUS	Dulag Luft	WV
	MC DANIEL PAUL E	T/Sgt	17159771	7/25/1944	AUS	Dulag Luft	MO

764	MC DONOUGH JOHN J	Sgt	32765717	10/4/1944	GER	Stalag Luft IV	NJ
	MC GARA DONALD R	2Lt	0-715094	8/7/1944	GER	Stalag Luft III	PA
	MC GRADY WILLIAM	S/SGT	18007034	8/22/1944	HUN	Stalag Luft IV	TX
	MC GUIRE JAMES G	S/Sgt	42021404	8/17/1944	RUM	Stalag Luft IV	NY
	MC KINNEY RALPH M	2Lt	0-687608	4/13/1944	HUN	Stalag Luft III	IL
765	MC LAIN WILLIAM L JR	2Lt	02057977	12/17/1944	GER	Stalag Luft I	AL
764	MEDELLIN EVEN G	S/Sgt	18202342	4/13/1944	NOR	Stalag Luft III	TX
	MERKOURIS THOMAS M	2Lt	0-824162	12/16/1944	ITA	Stalag Luft III	VA
	MERRY KENNETH E	F/O	T-127802	12/16/1944	ITA	Stalag Luft I	NC
	MERSINO JOHN F	Cpl	33675499	8/26/1944	YUG	Dulag Luft Hosp.	PA
764	MESSA FRANK	Sgt	31293316	10/4/1944	GER	Stalag Luft XII	RI
764	MILES VIRGIL S	2Lt	0-800743	5/10/1944	AUS	Dulag Luft	KS
765	MODROVSKY JOHN R	Sgt	32716898	12/17/1944	GER	Stalag Luft I	
765	MOHAN BERNARD J	2Lt	0-705759	8/3/1944	ITA	Dulag Luft	KS
	MOHLMAN ARTHUR W	1Lt	0-1052082	12/17/1944	GER	Stalag Luft XII	OH
	MORENO JOHN V	S/Sgt	20344522	4/13/1944	HUN	Stalag Luft III	MD
	MOORING ROBERT L	Sgt	38422557	12/17/1944	AUS	Dulag Luft West	TX
	MORLEDGE JACK C	Sgt	39294393	7/22/1944	RUM	Dulag Luft	CA
764	MORRELL DOUGLAS W		19048981	7/25/1944	RUM		CA
	MORRIS ROBERT D	Cpl	37731095	1/8/1945	YUG	Dulag Luft	MO
764	MORRISON DOUGLAS A	Cpl	14154460	4/25/1945	AUS	Dulag Luft West	GA
	MORT RICHARD A	S/Sgt	36459672	8/17/1944	RUM	Stalag Luft IV	MI

	MOSES JESSE F	2Lt	0-716810	12/18/1944	GER	Dulag Luft West	ID
	MUDGE WARREN R	2Lt	0-752769	5/27/1944	FRA	Undefined Code	CT
	MULLER WILLIAM J	2Lt	0-688385	5/24/1944	AUS	Undefined Code	KY
	MULLNER DAVID K	Cpl	39286629	8/26/1944	YUG	Stalag Luft IV	CA
	MUNDY CHARLES E	2Lt	0-2063458	12/16/1944	ITA	Stalag Luft III	FL
	MUNSON HOWARD E	2Lt	0-780316	12/17/1944	GER	Stalag Luft I	CA
764	MURPHY THOMAS I	Sgt	38172570	1/31/1945	AUS	Dulag Luft West	LA
	MYERS MICHAEL	T/Sgt	33082304	6/11/1944	RUM	Camp Chumen	PA
765	MYLLMAKI, EDWARD E		0-687623	7/15/1944			MA
765	NAGY, LOUIS W		32236429	2/21/1945			NJ
	NICOLAISON RUSSELL H	Sgt	37558581	12/17/1944	GER	Stalag Luft XII	MN
765	NIXON FORREST D JR	1Lt	0-440989	4/24/1944	ALB	Stalag Luft IV	AL
	NOBLE ROY L	Cpl	13190213	1/8/1945	YUG	Dulag Luft West	NY
765	NOESGES, THOMAS	2Lt	0-2056715	12/17/1944	GER	Dulag Luft Hosp.	IL
	NORRID BENJAMIN H	S/Sgt	34507445	5/27/1944	FRA	Stalag Luft IV	TN
	NORTON HAROLD R	Sgt	11085727	12/17/1944	GER	Stalag IIIA	VT
	NORTON ROBERT L JR	Sgt	34822745	7/25/1944	AUS	Dulag Luft	GA
	NOWOSILSKI HARRY	PFC	33788977	1/20/1945	AUS	Stalag 398	PA
764	OLSON DWIGHT B	2Lt	0-2057382	3/23/1945	AUS	Moosburg	MN
	OLSON ROLLAND T	2Lt	0-705781	7/25/1944	AUS	Stalag Luft I	MN
	OVIATT WINFIELD C	Cpl	12200537	12/17/1944	GER	Stalag Luft XII	NY
	PAGOTO JOSEPH	2Lt	0-709960	7/25/1944	GER	Stalag Luft I	MI

	PALMER JESSE H	T/Sgt	6934540	12/16/1944	ITA	Stalag Luft I	KS
764	PARADISE WILLIAM		0-775748	7/22/1944	RUM		NV
764	PAUL ALBERT I		33871548	9/1/1944	RUM		PA
765	PATTERSON WILLIAM E JR	1Lt	0-739050	7/25/1944	AUS	Stalag IVB	CA
765	PEEBLES JOHN A	Sgt	39621156	10/4/1944	GER	Stalag Luft IV	MT
765	PETERSON GEORGE C	T/Sgt	31100358	7/25/1944	AUS	Stalag Luft IV	ME
	POGORZELSKI DANIEL J	S/Sgt	32840090	8/7/1944	GER	Stalag Luft IV	NY
764	POWELL WILLIAM M	2Lt	0-823453	10/4/1944	GER	Stalag Luft I	CO
765	POWERS, DONNIE E		17128724	7/15/1944			KS
	PROUD DAVID E	S/Sgt	36458929	7/25/1944	GER	Stalag Luft IV	IL
765	QUALMAN THOMAS W	2Lt	0-2061046	12/17/1944	GER	Stalag Luft I	FL
	RADER JOSEPH G	Cpl	37497055	7/25/1944	AUS	Dulag Luft	MO
	RADOVSKY MILTON	1Lt	0-691800	7/25/1944	AUS	Stalag Luft I	NY
	RAINES ALVIN L	S/Sgt	37417737	5/27/1944	FRA	Dulag Luft	MO
765	RANIERI HARRY P	S/Sgt	32466582	5/31/1944	GER	Stalag IIA	NJ
	RAY HOWARD T	2Lt	0-722888	7/25/1944	AUS	Dulag Luft	TN
	REED GEORGE E	2Lt	0-692074	4/11/1944	HUN	Stalag Luft III	NY
	REEVE HAROLD I	Sgt	32368613	8/7/1944	GER	Stalag Luft IV	NJ
764	REYNOLDS JOHN F	S/Sgt	36270037	4/13/1944	HUN	Stalag VIIA	WI
765	REZNICEK DONALD T	Sgt	32765710	7/25/1944	AUS	Stalag Luft IV	NJ
765	RHODES DONALD J	1Lt	0-825981	4/11/1945	ITA	Dulag Luft West	NC
	RICCA PHILLIP	S/Sgt	39044893	8/24/1944	ITA	Stalag Luft IV	CA

764	RICE JOHN W	S/Sgt	32976665	3/23/1945	AUS	Dulag Luft West	NY
	ROBINSON DONALD H	Cpl	37446883	12/17/1944	GER	Stalag Luft I	IA
764	ROBINSON DONALD J	2Lt	0-2001251	1/31/1945	AUS	Dulag Luft West	PA
764	ROEBEN WALTER J	1Lt	0-685026	7/25/1944	GER	Stalag Luft III	CA
	ROSENCRANS G W JR	2Lt	0-704981	8/23/1944	AUS	Stalag Luft III	OH
	ROSS HARLAN P	2Lt	0-692078	4/2/1944	YUG	Stalag Luft I	NJ
765	ROSS TREFRY A	Sgt	39405340	12/17/1944	GER	Stalag Luft XII	AZ
764	RUDEL CHESTER H	2Lt	0-713554	12/17/1944	GER	Stalag Luft XII	
	RUTTER JAMES M	Cpl	39305204	12/17/1944	GER	Stalag Luft XII	OR
	RYAN TIMOTHY J JR	S/Sgt	31285389	7/30/1944	RUM	Romania 45-21	MA
764	RYDER WILLIAM	2Lt	0-752948	5/10/1944	AUS	Dulag Luft	CA
764	SALAS JOSE M	Sgt	38442085	7/25/1944	GER	Stalag 17B	
	SANDERS WILLIAM F	S/Sgt	33907008	3/23/1945	GER	Stalag Luft VIIA	MD
	SCHAEFFER EARL S	Cpl	33833119	7/25/1944	AUS	Stalag Luft IV	PA
764	SCHANK CARL R	S/Sgt	35278875	10/4/1944	GER	Stalag Luft IV	WV
	SCHIFFBAUER EDWARD D	S/Sgt	15019454	2/13/1945	AUS	Dulag Luft West	OH
	SCHONZEIT MARVIN	2Lt	0-668790	6/26/1944	AUS	Stalag Luft III	FL
	SCHWARTZ MURRAY	S/Sgt	12191951	6/26/1944	AUS	Stalag Luft IV	NY
765	SCHWEISBERGER MARTIN	2Lt	0-819627	6/25/1944	AUS	Stalag Luft III	IN
765	SCHWISOW LAUREN L	2Lt	0-772524	7/25/1944	AUS	Stalag Luft I	NE
764	SCOTT THEODORE R JR	2Lt	0-2065635	1/31/1945	AUS	Dulag Luft West	IL
	SERGIACOMI ARMAND E	Cpl	32709549	8/26/1944	YUG	Stalag Luft IV	NY

764	SEVERNS JACK R	2Lt	0-752944	4/13/1944	HUN	Stalag Luft III	CA
765	SHAFFER PAUL D	2Lt	0-771984	10/4/1944	GER	Stalag Luft I	PA
	SHEA LESTER W JR	Sgt	32448399	5/25/1944	GER	Stalag Luft III	NY
	SHERLOCK JOHN P	2Lt	0-711253	2/13/1945	AUS	Stalag Luft III	
	SIDOVAR NICHOLAS	2Lt	0-823998	12/17/1944	GER	Dulag Luft West	NJ
765	SIPPLE JAMES E	S/Sgt	13091570	7/25/1944	AUS	Stalag Luft IV	PA
765	SLAYTON LEONARD E	Sgt	38280283	4/24/1944	RUM	Stalag Luft III	TX
764	SLIGAR GLENN A	S/Sgt	39548073	3/23/1945	AUS	Dulag Luft West	NV
764	SMALL WILLIAM L	S/Sgt	11141038	4/25/1945	AUS	Dulag Luft West	MA
	SMITH ANDREW T	S/Sgt	6383767	7/12/1944	FRA	Dulag Luft	LA
764	SMITH CHARLES M JR	2Lt	0-683159	4/13/1944	GER	Stalag Luft III	ID
765	SMITH GERALD R	2Lt	0-824959	12/17/1944	GER	Stalag Luft I	CA
764	SMITH KENNETH B	2Lt	0-825700	12/17/1944	GER	Stalag Luft XII	IN
764	SMITH MARLIN R	S/Sgt	18009272	3/23/1945	AUS	Dulag Luft West	TX
765	SMITH WOODROW W	Sgt	35201035	10/4/1944	GER	Stalag Luft IV	WV
	SMYTH FREDERICK C	2Lt	0-2061068	12/17/1944	GER	Stalag IIIA	CO
	SNOW OLIVER I	Sgt	20148853	1/8/1945	YUG	Stalag Luft XII	ME
764	SPELIADES GEORGE A	Sgt	31409051	10/4/1944	GER	Stalag Luft IV	CT
	SPENCER LLOYD O JR	S/Sgt	37701960	2/13/1945	AUS	Dulag Luft West	CO
	SPRING MERRILL T	2Lt	0-695383	7/12/1944	FRA	Oflag VA	NY
	SQUIRES EDGAR A	Cpl	32982944	12/17/1944	GER	Dulag Luft Hosp.	NY
	STADALMAN CARL R JR	S/Sgt	17027803	5/18/1944	YUG	Stalag Luft III	KS

	STAUFFER, CLARENCE E	2Lt	0-695411	7/25/1944	AUS	Dulag Luft	FL
764	STEELE KAY B	2Lt	0-745972	4/13/1944	HUN	Stalag Luft III	ID
764	STEFENSON EDMUND F	1Lt	0-649279	5/31/1944	RUM	Dulag Luft	NY
765	STEINBERG CALVIN	T/Sgt	12187370	12/15/1944	AUS	Dulag Luft	TX
	STEVENSON THOMAS E	Sgt	33597030	12/17/1944	HUN	Dulag Luft	PA
765	STEWART CLIFTON J	Sgt	11091013	12/17/1944	GER	Dulag Luft West	MA
764	STITCH WRAY M	2Lt	0-700565	7/25/1944	GER	Dulag Luft	MN
	STOFTS BUREN H	S/Sgt	38131389	4/16/1944	YUG	Stalag Luft III	OK
765	STRINGER, GLENN E		17060928	7/15/.44			KS
	SULLENS OLIVER P JR	2Lt	0-717829	2/13/1945	AUS	Stalag Luft III	NM
	SULLIVAN JOSEPH A	1Lt	0-749799	7/25/1944	AUS	Dulag Luft	IL
764	SUNDQUIST VICTOR V	2Lt	0-713242	10/4/1944	GER	Stalag Luft I	CA
765	SUNDQUIST, GUS A		33702300	12/17/1944			PA
	SWINEHART ROBERT G	2Lt	0-816671	8/22/1944	YUG	Stalag Luft III	OH
764	TAYLOR JAMES G	1Lt	0-776125	1/31/1945	AUS	Dulag Luft West	MO
	TEN HAKEN MELVIN G	Cpl	36826670	12/16/1944	ITA	Stalag Luft I	WI
	THOMPSON ALBERT R	Sgt	39289201	8/22/1944	YUG	Stalag Luft IV	CA
	THOMPSON ROBERT K	2Lt	0-715407	8/7/1944	GER	Stalag VIIA	OH
765	TONEY, JOHN F		38588581	12/17/1944			OK
764	TOOTHMAN LAWRENCE R	2Lt	0-836681	4/25/1945	GER	Dulag Luft Hosp.	WA
	TRAYLOR CHARLIE T	Sgt	34652933	8/24/1944	ITA	Stalag Luft IV	SC
	TRIBBY JOHN F	S/Sgt	33689280	12/18/1944	GER	Dulag Luft West	PA

764	TROUP MERLE A	Sgt	6999672	4/13/1944	HUN	Stalag Luft III	PA
	TROUT JASPER D	S/Sgt	6293603	7/25/1944	AUS	Stalag Luft IV	TX
764	TRUMPY ROBERT T	Cpl	16186093	12/17/1944	GER	Stalag IXC	IL
765	TURGEON ARMAND R	Sgt	31109011	10/4/1944	GER	Stalag Luft III	NH
764	TURNER JOHN L JR	1Lt	0-696612	10/4/1944	GER	Stalag Luft I	NJ
	ULRICH EDWARD G	Sgt	12203084	4/2/1944	YUG	Stalag 17B	NJ
	VAITKUNAS WILLIAM R	S/Sgt	16080864	3/23/1945	GER	Stalag Luft 7A	IL
	VAN GUNDY DALE JR	PFC	16135216	10/16/1944	ITA	Stalag Luft IV	IL
765	VINSON RALPH D	S/Sgt	6951524	10/4/1944	GER	Stalag Luft IV	TX
765	VROOMAN, FRANCISCO L JR	2Lt	0-2059020	12/17/1944	GER	Stalag Luft I	CA
	WAGER HOWARD H	S/Sgt	32866006	8/24/1944	GER	Stalag Luft IV	NY
765	WAGGONER WILLIAM E	1Lt	0-680778	10/4/1944	GER	Stalag Luft I	AR
	WAKEFIELD R C	Cpl	14190769	12/16/1944	ITA	Stalag Luft I	AL
	WALKER HORACE A	2Lt	0-1692843	6/11/1944	RUM	Camp Chumen	FL
	WALLACE, IRVING G	S/Sgt	12024476	4/2/1944	YUG	Stalag 17B	NY
	WALSH JAMES P	S/Sgt	32463661	10/4/1944	GER	Stalag Luft XII	NJ
	WARD EDWARD P	Sgt	32766091	4/24/1944	RUM	Stalag Luft III	NJ
	WARD LEE P JR	1Lt	0-664480	12/16/1944	ITA	Stalag Luft III	TX
764	WARE ROBERT E	2Lt	0-723908	10/4/1944	GER	Stalag Luft I	KS
	WARREN ROBERT A JR	2Lt	0-797651	7/25/1944	AUS	Stalag Luft I	MA
765	WEEMS WILLIAM Z JR	2Lt	0-671716	7/30/1933	RUM	Camp Chumen	TX
	WELLS BRYAN	T/Sgt	17175058	7/25/1944	GER	Dulag Luft	MO
765	WENDTE DONALD D	S/Sgt	37476049	7/25/1944	AUS	Stalag Luft IV	NE

764	WENZLIK EDWARD T	2Lt	0-773028	3/23/1945	AUS	Dulag Luft West	CA
765	WEST SAMMY A	2Lt	0-752729	4/24/1944	GER	Stalag Luft III	TX
	WHITNEY EMIL L	2Lt	0-752730	4/2/1944	YUG	Stalag Luft I	ID
764	WIELAND ROY J	S/SGT	39716576	4/24/1945	AUS	Dulag Luft West	AZ
	WIEMANN NED W	S/Sgt	37553636	6/26/1944	AUS	Stalag Luft IV	MN
	WIEST WILLIAM H	T/Sgt	16076355	12/18/1944	GER	Stalag Luft I	IL
	WILD RUDOLPH H	S/Sgt	17159390	7/12/1944	FRA	Stalag Luft IV	IL
	WILEY THOMAS R	2Lt	0-721889	1/8/1945	YUG	Stalag Luft XII	TX
	WILHITE ROY B	Sgt	17097608	12/17/1944	HUN	Stalag Luft XII	IA
	WILSON JAMES P	S/Sgt	34601651	8/22/1944	YUG	Stalag Luft IV	NC
765	WILSON LESTER J	Cpl	38485586	7/25/1944	GER	Stalag Luft III	LA
	WILSON SIDNEY S	2Lt	0-795548	4/2/1944	YUG	Stalag Luft III	NY
	WINDHAM SAMUEL H	2Lt	0-771643	8/23/1944	GER	Stalag Luft III	TX
764	WINFIELD MARTIN D		11098596	8/23/1944			VT
765	WISEMAN NORMAN H	S/Sgt	35614386	7/25/1944	AUS	Stalag Luft IV	OH
	WOEHLKE ROBERT A	Cpl	33728213	12/17/1944	GER	Stalag Luft I	MD
	WREN JOHN R JR	2Lt	0-700594	8/24/1944	ITA	Stalag VIIA	MO
	YAW MYRON	Sgt	15334468	4/13/1944	HUN	Stalag Luft III	IN
	YOUNG BERNARD H	2Lt	0-708125	8/17/1944	RUM	Stalag Luft III	WI
764	ZEHFUSS EDWARD F	Sgt	13188915	1/31/1945	AUS	Dulag Luft West	PA
	ZINNER LEON	Sgt	12025667	5/27/1944	FRA	Stalag Luft IV	NY
	ZIPPILLI JOSEPH S	S/Sgt	32480166	4/13/1944	HUN	Stalag Luft III	NJ
	ZIVE SAMUEL	2Lt	0-1683594	6/26/1944	GER	Stalag Luft III	NJ

WAR CRIMES

THE RULES PROTECTING prisoners of war (POWs) are specific and were first detailed in the 1929 Geneva Convention. POWs must be treated humanely in all circumstances. They are protected against any act of violence, as well as against intimidation, insults, and public curiosity. International humanitarian law also defines minimum conditions of detention covering such issues as accommodation, food, clothing, hygiene and medical care.[1]

However many Allied crewmen, including members of the 461st Bomb Group, were certainly the victims of war crimes committed by Axis forces and civilians. Lt. Joshua R. Hooper and his crew took off on 21 February 1945 for a mission to Vienna, Austria. Just after bombs away he was forced to feather no. 1 engine due to a mechanical failure. His no. 4 engine also caught fire and was smoking badly. For a time he was able to follow the formation as it assembled after the bomb run. It eventually became obvious that he would not be able to keep up, or get his crew back to Italy. They decided the best bet would be to try for the Russian lines, but ran out of gas over Hungary. While the crew bailed out successfully, they were immediately picked up by Hungarian forces. After being questioned by three Hungarian officers they were turned over to the German commander of Budapest forces, SS-Brigadefuhrer Hellmuth Becker, commander of the 3rd SS Panzer Division Totenkopf (Deaths Head). They were seen in Becker's company that evening, but had disappeared by the morning. Local Hungarians asked the German soldiers what had happened to the Americans and were told they had been sent to another town. The Hungarians knew that was not true as the town had already fallen to the Russians. The crew was never recovered. With Russian forces closing in, Becker ordered the Deaths Head Division to cross the Danube River in an attempt to surrender to approaching American forces. However, the surrender was refused and

Becker and his entire division were turned over to the Russians. Becker escaped trial for his crimes, but was eventually executed in February 1953 by the Russians for his behavior while a prisoner. Lost with Lt. Hooper was Lt. Charles MacDonald, Co-Pilot; Lt. Frederick Bennehoff, Bombardier; Lt. W. Beach, Navigator; Sgt. Everett Taylor, Right Waist Gunner; Sgt. James Wolff, Left Waist Gunner; Sgt. Robert Schmitt, Top Turret; Sgt. D.M. Emmons, Ball Turret; Sgt. James De Lanzo, Tail Turret; and Sgt. James Gee, Nose Turret. Along with the Hooper crew Dr. Georg Hoffmann had identified 71 461st crewmen from 20 different crews who were either victims of assault or murder in Austria alone.

The people who committed war crimes came from all levels of society. They may have been guards in POW camps, or soldiers in combat. There were Gestapo and SS troops who killed airmen as soon as they were captured. However, the largest group, both in assaults and murders, was the local population. In some cases these were simple acts of revenge for the destruction of their towns by Allied bombers, but more were influenced by the Gestapo's control over the nation and the extensive Nazi propaganda machine portraying airmen not as airmen or pilots, but as "terror flyers" (Terrorflieger), "gangsters of the air" (Luftgangster), or "child murderers" (Kindermorder). Goebbels went to the extreme of ordering radio and print publications to eliminate the words "pilot", "aircrew" and "plane" from official vocabulary. Since Germans respected soldiers, Goebbels removed that status by claiming they were criminals who earned a bounty for each civilian killed. There was not only no sense of guilt associated with these crimes, but it was considered a patriotic act to participate in "Lynchjustiz" against Allied

murderers. Dr. Georg Hoffmann (see below) determined that there were approximately 1,500 assaults and murders in Austria alone. Oddly, there were more cases of mistreatment committed in Austria and Hungary than any other country.

German propaganda took the form of "exposing" the history and questionable culture of the United States, portraying American airmen as immoral savages responsible for the destruction of German cities. Lynching, or American justice, was considered appropriate because of the institutional racism against African-Americans and immoral American culture. Further, it gave German civilians an opportunity to fight their tormenters who were raining down misery, devastation, and death on the population.

There were two primary documents that promoted lynchjustiz. The first was an article written in May 1944 by Propaganda Minister Joseph Goebbels:

> *"It is only possible to protect the lives of downed enemy pilots with the use of armed forces, as the afflicted population would otherwise kill them. Who is right here? The murderers who, after their cowardly misdeeds, still expect humane treatment on the part of the victims or the victims who want to defend themselves according to the principle of an eye for an eye and a tooth for a tooth. The question is not difficult to answer. In any case, it would be intolerable to use German soldiers to protect child murderers from the rage-stricken parents who resort to self-defense after having just lost their most precious possessions through the brutal cynicism of the enemy."*

Head of the Nazi Party Chancellery Martin Bormann, also writing in May 1944, went further in admitting the local population had already been lynching downed flyers, said:

> *"In recent weeks, English and North American airmen have repeatedly shot at children playing, women and children working in the fields, peasants working on farms, wagons on the road, etc. during low-level staffing attacks. This is the most awful way to murder defenseless civilians – especially women and children. It has often occurred that downed*

crewmembers of such aircraft were immediately arrested and lynched on the spot by the outraged population. Police and criminal prosecution of the people involved were omitted."

Expanding on Bormann's letter, Albert Hoffman, National Defense Commissioner of the Gau Westphalia South, gave the following orders:

"Any fighter-bomber pilots shot down are on principle not to be protected from the indignation of the people. I expect from all police offices that they will refuse to lend their protection to these gangster types. Authorities acting in contradiction to the popular sentiment will be taken to account by me."

What had begun as individual acts of revenge had become a State sponsored policy for combating the enemy. The problem was the obvious fact that these policies violated international rules of law as they relate to POWs. Since arrival at the Luftwaffe reception camp granted automatic POW status, downed aircrews would be told they were not going to the reception camp, but were considered criminals that would be tried for their criminal acts. The exception to this policy was capture by the Wehrmacht or the Luftwaffe, who usually granted POW status according to the Geneva Convention. So many Allied airmen were murdered that a new word, which had nothing to do with Goebbels, became part of the vocabulary. That word was "Fliegermorde", meaning flyer murder.

Many of these criminals were tried in war crimes trials after the war. Others, however, escaped justice. While the search for these people continues today, many others would prefer that we forget the past. To their credit the government of Austria has chosen to present the crimes that occurred on Austrian soil to the world. An eight-year project conducted by Austrian researchers Dr. Georg Hoffmann and Dr. Nicole-Melanie Goll resulted in the book "Missing in Action: Failed to Return", which the Austrian government posted on-line for anyone to see. The remarks below were given by Dr. Hoffmann at a ceremony in Vienna on April 22, 2016 when the commemoration book was presented to the U.S. Ambassador to Austria Alexa Wesner.

Honoring the War Dead: Austria Says Thank You to WWII Allied Airmen

by Georg Hoffmann, PhD.

The cultural scientist Aleida Assmann has named – within her studies about memory – 70 years after the end of an occurrence as an important time mark: a point in time where a communicative/collective memory is transforming into a cultural one. Or to say it in other words: where we won't be able any more to learn from contemporary witnesses; where societies have to decide which elements should be integrated into an active commemoration to give future generations the chance to learn from them. It is a decision as to which elements should be preserved and handed over to the next generations.

We are now standing at this point in time – when we are looking on the Second World War and the time of the Nazi regime. Personal memories about this time, about suffering, about violence are disappearing, traces are vanishing, names are forgotten, and remembrance of those who are not with us any more is fading.

So what does this point in time mean for us and for the goal to learn from history? How can the knowledge of the traumatic and painful experiences of Nazi terror, war and Holocaust be preserved in memory? And how could remembrance of all those be preserved who gave their lives in fighting the Nazi regime?

For fellow Austrian historian Nicole-Melanie Goll and I, who have spent the last eight years compiling a database of all Allied air crashes over Austria during World War II, part of our initiative has been to create a memory book **"Missing in Action: Failed to Return."** This book, now posted online by the Austrian government, contains the names of American and British soldiers, who have not before been part of an official commemoration in Austria – who had been, and I want to say it that clear: forgotten.

These soldiers were pilots and airmen of the U.S. Army Air Force, the Royal Air Force, the South African Air Force, The Royal Australian, Canadian and New Zealand Air Force and who lost their lives in air war over what is today Austria. They were very young men, who fought in a war – far away from their hometowns and families – in order to liberate Europe and therefore also Austria from a tyranny and dictatorship.

They lost their lives not only in air battles and crashes of their planes. They also became victims of war crimes and violence, which the Nazi regime set loose under the name of "lynch law" and as a revenge for Allied air attacks. Inside the whole German Reich and therefore also inside Austria the so -called "terrorflyers" were hunted.

"Terrorflyer" was the catchphrase of Nazi propaganda for Allied crews. Airmen who were captured were humiliated, mistreated and also murdered in public. The Nazi state especially aimed to take away their identity – by stealing and extinguishing their names! Funerals for airmen were forbidden, false names were put on graves, bodies were carried away and measures were taken so that nobody ever would be able to identify them

The stigmatization of "terrorflyers" or "Luftgangster" was a very sustainable one.

The memory book now has the goal to call the names of these men into memory and to document their fate, which was in so many cases unknown. The book contains 1,715 names from airmen who died in Austria – and it gives the exact time and place of their death. Among them are 1,582 Americans. It also contains the names of 113 who are still Missing in Action. Their bodies were never found, and a search for them is still going on – e.g. carried out by institutions of the U.S. Department of Defense. The book is giving the newest and best information available about the whereabouts of these men.

Besides the listing of names, the book points out the fates of several individuals that are especially moving. Like the story of Walter Manning, who was shot down on April 1, 1945, near Linz, heavily tortured after his capture and hanged publicly at the airbase of Linz-Hoersching three days later. His case was never solved, and Manning himself is totally forgotten. Or the fate of a group of American airmen around LeRoy Teschendorf who were taken away by SS men and murdered – on order of the August Eigruber – inside the Concentration Camp Mauthausen. The whole group disappeared without a trace.

Last Name	First Name	Rank	MOS	Sq.	Date/Place of Death	Aircraft Number	Day/Place/Region of Crash
Albregts	August H.	2nd Lt	P	765	1/20/1945 Linz (Franckviertel)	42-51378	1/20/1945 Linz (Franckviertel) Oberösterreich
Baird	William R.	1st Lt	P	764	3/23/1945 Schönfeld im Marchfeld	44-41091	3/23/1945 Schönfeld im Marchfeld Niederösterreich
Carlock	Alfred F	T/Sgt	RWG	766	6/26/1944 Dürnkrut	41-28679	6/26/1944 Dürnkrut Niederösterreich
Thomas	Perry J.	T/Sgt	BTG	766	6/26/1944 Dürnkrut	41-28679	6/26/1944 Dürnkrut Niederösterreich
Turner	Robert M.	Sgt	TTG	765	6/26/1944 Dürnkrut	41-28679	6/26/1944 Dürnkrut Niederösterreich
Fulcher	Yancey L.	T/Sgt	LWG	764	7/25/1944 Zauchersteg	41-28693	7/25/1944 Zauchersteg Niederösterreich
Hagen	James P.	T/Sgt	RWG	765	7/25/1944 Zauchersteg	41-28693	7/25/1944 Zauchersteg Niederösterreich
Jelens	William	S/Sgt	BTG	767	7/25/1944 Hofkirchen im Traunkreis	41-28836	7/25/1944 Hofkirchen im Traunkreis Oberösterreich
Lucera	Anthony A.	S/Sgt	NG	767	7/25/1944 Hofkirchen im Traunkreis	41-28836	7/25/1944 Hofkirchen im Traunkreis Oberösterreich
Mancourt	Edward M.	2nd Lt	B	767	7/25/1944 Hofkirchen im Traunkreis	41-28836	7/25/1944 Hofkirchen im Traunkreis Oberösterreich
Hesser	Joseph B.	1st Lt	P	766	7/25/1944 Winden	41-28850	7/25/1944 Winden Oberösterreich
Kimble	Robert H.	S/Sgt	RWG	766	7/25/1944 Winden	41-28850	7/25/1944 Winden Oberösterreich
Tyler	George T.	Cpl	TG	766	7/25/1944 Winden	41-28850	7/25/1944 Winden Oberösterreich
Westfall	Coy	S/Sgt	TTG	766	7/25/1944 Winden	41-28850	7/25/1944 Winden Oberösterreich
Myers	Glen A.	Cpl	TTG	765	7/25/1944 Luftenberg	41-29529	7/25/1944 Luftenberg Oberösterreich
Gibbs	Robert H.	Cpl	NG	765	1/20/1945 Linz (Franckviertel)	42-51378	1/20/1945 Linz (Franckviertel) Oberösterreich
Harris	Arthur M.	Cpl	TTG	765	1/20/1945 Linz (Franckviertel)	42-51378	1/20/1945 Linz (Franckviertel) Oberösterreich
Helwig	Fred A.	Cpl	TG	765	1/20/1945 Linz (Franckviertel)	42-51378	1/20/1945 Linz (Franckviertel) Oberösterreich
Kenyon	Lloyd E.	S/Sgt	P	764	1/20/1945 Linz (Franckviertel)	42-51378	1/20/1945 Linz (Franckviertel) Oberösterreich
Krueger	John H.	F/O	B	765	1/20/1945 Linz (Franckviertel)	42-51378	1/20/1945 Linz (Franckviertel) Oberösterreich

McRae	George L.	Cpl	LWG	765	1/20/1945 Linz (Franckviertel)	42-51378	1/20/1945 Linz (Franckviertel) Oberösterreich
Mitchell	Grover F.	1st Lt	P	764	1/20/1945 Linz (Franckviertel)	42-51378	1/20/1945 Linz (Franckviertel) Oberösterreich
Monaco	Anthony J.	Cpl	BTG	765	1/20/1945 Linz (Franckviertel)	42-51378	1/20/1945 Linz (Franckviertel) Oberösterreich
Morrell	Joseph W.	Cpl	RWG	765	1/20/1945 Linz (Franckviertel)	42-51378	1/20/1945 Linz (Franckviertel) Oberösterreich
Raimo	George	F/O	N	765	1/20/1945 Linz (Franckviertel)	42-51378	1/20/1945 Linz (Franckviertel) Oberösterreich
DiMartino	Louis	S/Sgt	LWG	765	7/25/1944 Stranzberg	42-52368	7/25/1944 Stranzberg Oberösterreich
Duzan	Fred C.	S/Sgt	BTG	765	7/25/1944 Stranzberg	42-52368	7/25/1944 Stranzberg Oberösterreich
First	Harry M.	1st Lt	N	765	7/25/1944 Stranzberg	42-52368	7/25/1944 Stranzberg Oberösterreich
Fulks	Glennial	1st Lt	P	765	7/25/1944 Stranzberg	42-52368	7/25/1944 Stranzberg Oberösterreich
Jaske	Walter R.	S/Sgt	NG	765	7/25/1944 Stranzberg	42-52368	7/25/1944 Stranzberg Oberösterreich
Krivitsky	Morris	S/Sgt	TTG	765	7/25/1944 Stranzberg	42-52368	7/25/1944 Stranzberg Oberösterreich
Steward	Robert B.	F/O	B	765	7/25/1944 Stranzberg	42-52368	7/25/1944 Stranzberg Oberösterreich
Bigelow	Robert L.	1st Lt	P	766	5/24/1944 Sigleß	42-52405	5/24/1944 Sigleß Burgenland
Dickinson	John W.	Cpt	CP	766	5/24/1944 Sigleß	42-52405	5/24/1944 Sigleß Burgenland
Fisher	Floyd E.	2nd Lt	N	766	5/24/1944 Sigleß	42-52405	5/24/1944 Sigleß Burgenland
Lane	Robert G.	S/Sgt	TG	766	5/24/1944 Sigleß	42-52405	5/24/1944 Sigleß Burgenland
Paccione	Michael T	S/Sgt	LWG	766	5/24/1944 Sigleß	42-52405	5/24/1944 Sigleß Burgenland
Rudisill	Edwin A.	S/Sgt	TTG	766	5/24/1944 Sigleß	42-52405	5/24/1944 Sigleß Burgenland
Tarpey	Joseph P.	T/Sgt	RWG	766	5/24/1944 Sigleß	42-52405	5/24/1944 Sigleß Burgenland
Lavery	Clifford J.	T/Sgt	LWG	767	7/25/1944 Breitenau	42-52451	7/25/1944 Breitenau Oberösterreich
Roberts	Earl R.	S/Sgt	BG	767	7/25/1944 Breitenau	42-52451	7/25/1944 Breitenau Oberösterreich
Rogan	Frank G.	S/Sgt	RWG	767	7/25/1944 Molln	42-52451	7/25/1944 Breitenau Oberösterreich
Albright	William M.	F/O	B	764	7/25/1944 Bad Zell	42-52459	7/25/1944 Riegl Oberösterreich

Dahlke	Gene C.	Sgt	RWG	764	7/25/1944 Riegl	42-52459	7/25/1944 Riegl Oberösterreich
Dalbec	Joseph R.	Sgt	LWG	764	7/25/1944 Riegl	42-52459	7/25/1944 Riegl Oberösterreich
Ducray	Julian R.	Cpl	BTG	764	7/25/1944 Riegl	42-52459	7/25/1944 Riegl Oberösterreich
Harris	James E.	1st Lt	CP	767	7/25/1944 Riegl	42-52459	7/25/1944 Riegl Oberösterreich
Towne	Henry S.	2nd Lt	CP	764	5/10/1944 Apetlon	42-78212	5/10/1944 Apetlon Burgenland
Wallace	William C.	1st Lt	P	764	10.5.1944 Apetlon	42-78212	10.5.1944 Apetlon Burgenland
Ellis	Herbert	Sgt	BTG	767	7/25/1944 Erdleiten	42-94732	7/25/1944 Erdleiten Oberösterreich
Farrell	Edward F.	Sgt	RWG	767	7/25/1944 Erdleiten	42-94732	7/25/1944 Erdleiten Oberösterreich
Hanson	Arne P.	2nd Lt	N	767	7/25/1944 Erdleiten	42-94732	7/25/1944 Erdleiten Oberösterreich
Moreau	Norbert L.	S/Sgt	TG	767	7/25/1944 Erdleiten	42-94732	7/25/1944 Erdleiten Oberösterreich
Peo	Robert	Cpl	NG	767	7/25/1944 Erdleiten	42-94732	7/25/1944 Erdleiten Oberösterreich
Brown	Robert S	1st Lt	N	764	7/25/1944 Konzentrationslager Mauthausen	42-95383	7/25/1944 Hargelsberg Oberösterreich
Harrell	Lynwood C.	S/Sgt	NG	764	7/25/1944 Konzentrationslager Gusen	42-95383	7/25/1944 Hargelsberg Oberösterreich
Hartford	Herbert H.	2nd Lt	P	764	7/25/1944 Konzentrationslager Mauthausen	42-95383	7/25/1944 Hargelsberg Oberösterreich
Herbert	Ira W.	T/Sgt	LWG	764	7/25/1944 Hargelsberg	42-95383	7/25/1944 Hargelsberg Oberösterreich
Kennedy	John L.	Sgt	TTG	764	7/25/1944 Konzentrationslager Mauthausen	42-95383	7/25/1944 Hargelsberg Oberösterreich
Robinson	Donald D.	S/Sgt	757	764	7/25/1944 Hargelsberg	42-95383	7/25/1944 Hargelsberg Oberösterreich
Jones	Lawton M.	2nd Lt	CP	764	10/4/1944 Mining	44-41039	10/4/1944 Mining Oberösterreich
Ellis	Thomas R	Cpl	TTG	766	1/20/1945 Linz (Römerberg)	44-48761	1/20/1945 Linz (Römerberg) Oberösterreich
Keenan	Francis T.	Sgt	LWG	766	1/20/1945 Linz (Franckviertel)	44-48761	1/20/1945 Linz (Franckviertel) Oberösterreich

Knoke	Norman E.	2nd Lt	N	766	1/20/1945 Linz (Römerberg)	44-48761	1/20/1945 Linz (Römerberg) Oberösterreich
Mitchell	Burton F.	Cpl	TG	766	1/20/1945 Linz (Römerberg)	44-48761	1/20/1945 Linz (Römerberg) Oberösterreich
O'Neal	Joseph M.	2nd Lt	P	766	1/20/1945 Linz (Römerberg)	44-48761	1/20/1945 Linz (Römerberg) Oberösterreich
Rossini	Gino F.	Cpl	NG	766	1/20/1945 Linz (Römerberg)	44-48761	1/20/1945 Linz (Römerberg) Oberösterreich
Rothe	Roderick E.	2nd Lt	P	766	1/20/1945 Linz (Römerberg)	44-48761	1/20/1945 Linz (Römerberg) Oberösterreich
Teller	Donald S.	1st Lt	B	766	1/20/1945 Linz (Römerberg)	44-48761	1/20/1945 Linz (Römerberg) Oberösterreich
Jones	William	2nd Lt	CP	764	4/25/1945 St. Georgen am Walde	44-49511	4/25/1945 St. Georgen am Walde Niederösterreich
Spehalski	Raymond E.	2nd Lt	P	764	3/26/1945 Kindberg	44-49641	3/26/1945 Kindberg Steiermark

But the book also is pointing out fates of pilots, who – like Charles Faxon – sacrificed their lives in order to save their crew. Or like the two U.S. airmen and very close friends Robert Brewer and Robert Spicer, who tried to jump out of their burning plane together with the last remaining chute. Brewer did not survive this jump, and Spicer was shot several hours later by a firing squad of the SS police near Bleiburg in southern Austria. Behind these names one can also find attempts of the population to help downed airmen, which in many cases lead to the death of both the airmen and his helper. Or the fate of Jewish airmen, like Manfred Grunberg, who were secluded from their comrades after capture and often shot immediately. The memory book now has the goal to give back the names to these Allied soldiers who lost their lives in fighting Nazi Germany and to add them to the names of all people to whom Austria and all Austrians owe their freedom. The book especially addresses the families and relatives of these airmen, giving them the best obtainable information about the whereabouts of their loved ones. It also has the goal to support the search for still missing soldiers. [2]

The book is posted online at http://www. bundesheer.at/ download_archiv/pdfs/ missing_in_ action.pdf

[1]International Committee of the Red Cross Article 29 October 2010

https://www.icrc.org/en/document/protected-persons/prisoners-war

[2]Liberaider Magazine, June 2017, pp 26-27

(3)

MACHINES

461st Bomb Group Aircraft

EVERY SHIP of the 461st had its own story. The original aircraft were olive drab "H" models produced by the Ft Worth, Douglas Tulsa, and Ford Willow Run plants. Of the 62 aircraft who left the United States to form the core of the 461st Bombardment Group (H), 29 would crash, 12 would transfer to other units, 10 would be salvaged after battle damage, 2 crash landed, 2 ditched, 2 were involved in a mid-air collision on the group's first mission, 1 was condemned, 1 salvaged non battle damage, and three survived to return home after the war.

Some aircraft were olive drab while others had a natural metal finish. Some carried radar and were called Pathfinders, while some of the Pathfinders were painted dull gray overall and titled Lone Wolves.

B-24H-10-DT 41-28670, (767th Sq., #63)

41-28670

Malfunction-Sired by Ford, an original aircraft of the 461st, departed the US with the Walter's crew on 1 January 1944 and arrived in Italy February 3rd. Salvaged due to non-combat damage on 16 April 1944.

B-24H-10-DT 41-28679, (766th Sq., #47)

41-28679

Heaven Can Wait, left the US with the Holmes crew on 25 January 1944, and arrived in Italy 2 February. Lost 26 June 1944 with Lt. Zive at the controls. Losing power in two engines, he dropped his load before reaching the target, aborted the mission and was last seen being attacked by fighters (MACR 6399).

B-24H-10-DT 41-28680,
(766th Sq., #42)

Arriving in Italy 1 February 1944 with the Ahlberg crew, it was transferred to the 47th Bomb Wing 28 February 1944 and lost flying with the 451st Bomb Group on 23 April 1944.

B-24H-10-DT 41-28681, (765th Sq., #23)

41-28681

Our Baby arrived 9 February 1944 with Lt. Taylor at the controls. Transferred to the 767th as #62 and later #73. Lost for unknown reasons on 22 August 1944 over Lubau, Austria. (10 POW).

B-24H-10-DT 41-28683, (767th Sq., #69)

41-28683

Arrived 2 March 1944 and lost in a taxiing accident on 12 March 1944 when the main gear collapsed after the aircraft slipped off the runway into a draining ditch.

B-24H-10-DT 41-28685, (765th Sq., #24)

41-28685

Leading Lady Arrived 9 February 1944 with Marion Mixson at the controls. Lost to flak over Ploesti on 31 May 1944 MACR 6126 (1 KIA, 9 POW).

B-24H-10-DT 41-28689, (766th Sq., #40)

Sweet Chariot, Arrived 19 February 1944, lost 23 August 1944 over Markersdorf from engine trouble and fighters. MACR 8766 (1 KIA, 9 POW).

B-24H-10-DT 41-28693, (764th Sq., #1)

41-28693

War Eagle, arrived 14 February 1944, lost 25 July 1944 over Linz to enemy fighters. MACR 11977 (2 KIA, 8 POW).

B-24H-10-DT 41-28708, (766th Sq., #55)

41-28708

The Mountaineer, arrived 31 January 1944, transferred to the 450th Bomb Group. Crashed with that group 16 August 1944.

B-24H-10-DT 41-28717, (765th Sq., #30)

41-28717

Big Stinky, arrived 22 February 1944. Hit by flak on the 24 April 1944 mission to Bucharest, Lt. Nixon was able to fly as far as the coast of Albania before being forced to bail out the crew. MACR 4627 (3 KIA, 7 POW).

B-24H-10-DT 41-28724, (767th Sq., #71)

41-28724

Jizzy Outch, arrived 2 February 1944, lost to flak over Bad Voslau Airdrome on 23 April 1944.

B-24H-10-DT 41-28725, (765th Sq., #27)

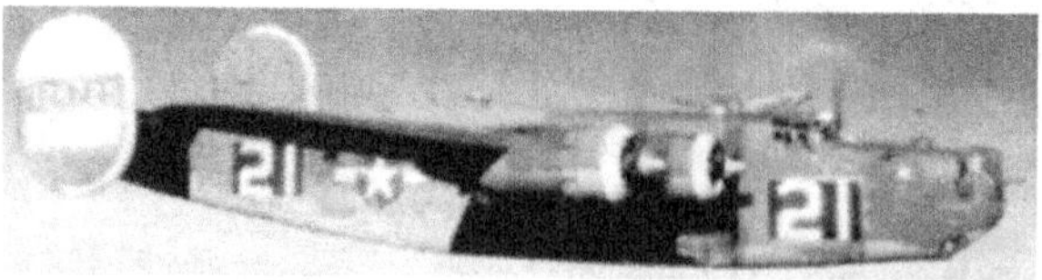

41-28725

Invictus, arrived in Italy 8 February 1944, salvaged (non-combat) on 4 August 1944.

B-24H-10-DT 41-28726, (764th Sq., #6)

41-28726

Mister Period (.), arrived 13 February 1944, crash landed returning from 10 June 1944 mission.

B-24H-10-DT 41-28732, (764th Sq., #18)

41-28732

Swee' Pea, arrived 7 February 1944. On the 25 July 1944 mission to Linz the group was attacked by 150 German planes. Eleven ships went down over the target and four more, including Swee' Pea were lost trying to return to the base. They tried to use parachutes to replace the flaps, but ran off the runway and crashed.

B-24H-10-DT 41-28734, (764th Sq., #15)

41-28734

Nov Schmoz Ka Pop, arrived 10 February 1944. While attacking Genoa, Italy on 12 August 1944 it was hit by flak causing a fire. Without warning the plane exploded. MACR 16376 (10 KIA).

B-24H-10-DT 41-28737, (765th Sq., #20)

41-28737

Rhode Island Red, arrived 8 February 1944, salvaged 15 August 1944.

B-24H-10-DT 41-28740, (765th Sq., #36)

41-28740

The Bat, arrived 6 February 1944. Transferred to the 451st Bomb Group on 28 February 1944.

B-24H-15-DT 41-28836, (767th Sq., #69)

This ex-484th Bomb Group aircraft arrived on 29 March 1944. Lost on 25 July 1944 mission against the Hermann Goering Tank Works, one of 15 planes either shot down by German fighters over the target or crash landing at friendly bases while trying to get home. MACR 7038 (3 KIA, 7 POW).

PB-24H-15-DT 41-28850, (766th Sq., #44)

This pathfinder aircraft arrived on 16 April 1944. Lost 25 July 1994 over Linz, Germany. The pilot, Lt. Hesser, and three others were killed. Six others became POWs.

PB-24H-15-DT 41-28856, (764th Sq.)

Pathfinder. Arrived 5 June 1944. Condemned following a crash-landing 18 July 1944.

PB-24H-15-DT 41-28858, (767th Sq., #62)

41-28858

Lady Edith, Pathfinder. Arrived 11 June 1944, condemned 14 April 1945.

B-24H-15-DT 41-28867, (765th Sq., #21)

41-28867

Sleepy Time Gal, Arrived 2 April 1944. Shot down by fighters over Linz, Germany 25 July 1944. MACR 7113 (10 POW).

PB-24H-15-DT 41-28891, (765th Sq., #25)

41-28891

Billie K, Arrived 16 April 1944. Returned to the US on 9 June

1945.PB-24H-15-DT 41-28903, (764th/765th Sq., #16)

41-28903

Myra G, Pathfinder, arrived 9 June 1944, condemned following a crash on 15 April 1945. (Photo shows Myra G being converted to a Lone Wolf aircraft.

B-24H-15-DT41-28913, (765th Sq., #67)

41-28913

Arrived 9 April 1944, aircraft was severely damaged during the aerial battle on 17 December 1944. The pilot, Lt, Helton decided to try reaching friendly lines while Sgt Piccoli, although wounded, tried to smother a fire in the tail with his body. Unable to make friendly lines Lt. Helton crash landed it and the crew became POWs.

B-24-15-DT 41-28927, (765th Sq., #22)

41-28927

Arrived 31 March 1944, Salvaged 8 May 1944.

B-24H-15-DT 41-28940, (765th Sq., #36)

41-28940

Mary L, who arrived on 20 April 1944, crashed 12 November returning from a test flight.

B-24H-20-DT 41-28970, (765th Sq., #33)

41-28970

Liberty Belle, Arrived 17 April 1944, crashed as a result of mechanical failure 3 August 1944. (4 POWs, 6 Evaded).

B-24H-20-DT 41-29001

Arrived 30 April 1944, condemned following crash-landing returning from 7 July 1944 mission to Blechhammer.

B-24H-20-DT 41-29006, (765th Sq.)

Arrived 20 April 1944, ditched in the Adriatic 20 November 1944 returning from a mission to Blechhammer. The pilot, Charles Kahn, and three others were lost. Six others were rescued.

B-24H-10-CF 41-29268, (766th Sq., #52)

Tou Jour Gay, Arrived 31 January 1944, salvaged 22 June 1944.

41-29289

The Plastered Bastard, Arrived 2 March 1944. Following a landing accident the pilot exclaimed "Well, we really plastered this bastard." The name stuck. Salvaged 27 July 1945.

B-24H-10-CF 41-29289, (764th Sq., #5)

41-29289

Battle Crate, Arrived 7 February 1944. Lost 3 July 1944 over Yugoslavia due to mechanical failure. The navigator died when his chute did not open. The other nine crew members were picked up by either Partisans or Chetniks and returned to allied control.

B-24H-10-CF 41-29313, (766th Sq., #45)

41-29321

Hottest ??? In Town, Arrived 31 January 1944. When a commanding officer objected to the nickname, the offensive word was replaced by three question marks. Hard hit by flak over Ploesti on 22 Jul 1944, Lt. Holmes was able to get his ship back, but was forced to bail out his crew near the field.

B-24H-10-CF 41-29321, (767th Sq., #64)

41-29321

Flown over by the Strong crew, it arrived on 2 February 1944. It returned to the US on 9 June 1945, one of the few original aircraft to survive the war.

B-24H-10-CF 41-29325, (765th Sq. #29)

41-29325

Hare Power, Arrived 14 February 1944. Transferred to the 454th Bomb Group on 28 February 1944.

B-24H-10-CF 41-29332, (767th Sq., #67)

Transferred to the 47th Bomb Wing on 28 February 1944.

B-24H-10-CF 41-29333, (764th Sq., #3)

41-29333

Stinky. This original aircraft arrived 10 February 1944. Salvaged following the 5 November 1944 mission to the Ali Pasin Most Marshalling Yard in Sarajevo, Yugoslavia.

B-24H-10-CF 41-29334, (764th Sq., #11)

41-29334

Ol' Bird, Arrived 13 February 1944, salvaged (non-combat) on 20 June 1945.

B-24H-10-CF 41-29335, (764th Sq., #9)

41-29335

Evil Weevil, arrived 2 March 1944, salvaged (non-combat) on 11 February 1945.

B-24H-15-CF 41-29336, (766th Sq., #50)

Arrived 6 February 1944, lost 2 April 1944 when involved in a mid-air collision on the group's first mission of the war. MACR 4087.

B-24H-15-CF 41-29337, (766th Sq., #53)

41-29337

Dwatted Wabbit, Original aircraft arrived 31 January 1944. Trying to return from 12 July 1944 mission to Nimes, France it was forced to ditch in the Mediterranean MACR 6808 (10 KIA).

B-24H-15-CF 41-29338, (764th Sq., #10)

Scrounch, Arrived 16 February 1944, transferred to 450th Bomb Group and lost with that group 25 April 1944.

B-24H-15-CF 41-29341, (767th Sq., #76)

Wheels of Justice, Arrived 2 February 1944, crashed 20 April 1944.

B-24H-15-CF 41-29362, (764th Sq., #7)

41-29362

Lucky Seven, Arrived 19 February, lost over Ploesti on 22 July 1944 MACR (1 KIA, 9 POW).

B-24H-15-CF 41-29367, (765th Sq., #32)

Lady Bird, Arrived 19 February 1944. Caught fire the night of 20 May 1944 due to a short in the hydraulic pressure switch. Fully loaded with bombs the plane exploded destroying two other ships and damaging three more.

B-24H-15-CF 41-29517 (766th Sq., #32)

41-29517

Pisces, originally assigned to the 486th BG, 834th BS, 8th Air Force, it would be transferred to the 461st. Condemned 19 June 1945.

B-24H-15-CF 41-29519, (767th Sq.)

Originally assigned to the 484th BG, it was transferred to the 461st and salvaged (non-combat) on 9 June 1944.

B-24H-15-CF 41-29529, (765th Sq., #23)

Hard Guy, originally a 484th BG aircraft, it was transferred to the 461st and lost on 25 July 1944 MACR 9844 (10 POW).

41-29529

B-24H-15-CF 41-29564, (Squadron Unknown)

Transferred from the 486th Bomb Group, 8th Air Force, Condemned, salvaged on 30 April 1944.

B-24H-20-CF 42-50303, (765th Sq., #22½)

42-50303

Our Hobby, Arrived 30 April 1944, lost 15 July 1944 over Ploesti when hit by flak. MACR 6962 (10 POW).

B-24H-25-CF 42-50397, (765th Sq.)

Arrived 22 May 1944. Severely damaged 22 July 1944 by flak over Ploesti the ship, flown by Col. Glantzberg and Lt. Specht, was able to make it back to base, but was declared Class 26 and salvaged.

PB-24J-1-FO 42-50605, (766th Sq., #50)

42-50605

Lo and Behold, Arrived 11 July 1944, returned to the US on 19 June 1945.

B-24J-1-FO 42-50613, (764th Sq., #5)

42-50613

Jo, an original 8th Air Force aircraft, it returned to the US on 7 June 1945.

B-24J-5-FO 42-50802, (764th Sq.)

Arrived 21 July 1944, involved in a taxi accident with 41-28730 on 28 July 1944 and condemned.

B-24J-5-FO 42-50948, (766th Sq.)

Arrived 14 July 1944, lost over Ploesti on 22 July 1944 MACR 7005 (10 POW).

B-24J-5-FO 42-50953, (765th Sq., #39)

41-50953

The Flying Finger, Arrived 21 July 1944, lost to German fighters during Odertal mission 17 December 1944 MACR 10677 (3 KIA, 7 POW).

B-24J-5-FO 42-50970, (764th Sq., #13)

42-50970

Luck Lady, Transfer from the 484th Bomb Group, lost 4 October 1944 when hit by flak over Munich. MACR 8975 (10 POW).

B-24J-5-DT 42-51306, (Squadron Unknown)

Arrived 6 August 1944, transferred to the 451st Bomb Group and lost with that group on 2 December 1944.

B-24J-5-DT 42-51319, (767th Sq., #69)

42-51319

Welcome Wagon, Arrived 23 July 1944, lost to German fighters during Odertal mission 17 December 1944. MACR 10651 (3 KIA, 7 POW).

B-24J-5-DT 42-51322, (767th Sq., #66)

Arrived 27 July 1944, lost to German fighters during Odertal mission 17 December 1944 MACR 10742 (7 POW, 3 evadees).

B-24J-5-DT 42-51324, (764th/765th Sq., #12)

42-51338

10 Men Back/Paulette, Arrived 4 August 1944, lost to German fighters during Odertal mission 17 December 1944 MACR 10682 (1 KIA, 9 POW).

B-24J-5-DT 42-51336, (766th Sq., #46)

Shady Lady, Arrived 27 July 1944, lost to fighters on 24 August 1944 MACR 8327 (10 POW).

B-24J-5-DT 42-51338, (766th Sq., #54)

42-51338

Arrived 26 August 1944, lost to German fighters during Odertal mission 17 December 1944 MACR 8932 (10 POW).

B-24J-5-DT 42-51346, (765th Sq., #31)

Jake's Neighbor, Arrived 6 August 1944, developed engine trouble while on mission to Linz. All 10 crew members successfully bailed out, but 4 drowned when they landed in a flooded area. The other six became POWs. MACR 10676.

B-24J-5-DT 42-51378, (765th Sq., #25)

42-51378

Arrived 17 September 1944, lost to flak over Linz on 20 January 1945. MACR 11806 (10 KIA).

B-24J-5-DT 42-51388, (Squadron Unknown)

Arrived 7 March 1945, returned to the US on 28 June 1945.

B-24J-10-DT 42-51406, (764th Sq., #11)

42-51406

Arrived 24 September 1944, crashed 7 February 1945.

B-24J-10-DT 42-51412, (764th Sq.)

Arrived 21 August 1944, lost due to engine trouble over Hungary on 31 January 1945. (10 POW).

B-24J-10-DT 42-51422, (767th Sq., #72)

Arrived 8 September 1944, crashed 20 November 1944.

B-24J-10-DT 42-51430, (765th Sq., #24)

42-51430

The Tulsamerican, Arrived 27 September 1944, damaged by fighters 17 December 1944, crashed into Adriatic Sea attempting to reach the Isle of Vis MACR 16507 (3 KIA, 7 rescued).

B-24J-5-FO 42-51474, (765th Sq.)

Arrived 4 August 1944, returned to the US on 12 June 1945.

B-24J-5-FO 42-51501, (767th Sq., #69)

42-51501

Zombie of 69 The Second, Former 8th Air Force aircraft transferred to the 461st on 19 December 1944, salvaged (battle damage) 15 March 1945.

B-24J-5-FO 42-51599, (765th Sq., #34)

Transfer from the 8th Air Force on 19 December 1944. Returned to the US on 12 June 1945.

42-51599

B-24J-5-FO 42-51606, (764th Sq., #9)

Iggy, Arrived 10 August 1944, salvaged (battle damage) 8 October 1944.

42-51606

B-24J-5-FO 42-51610, (765th Sq., #33

Marjorie H, Arrived 11 October 1944, shot down by fighters over Vienna on 21 February. MACR 12473 (12 POW).

B-24J-10-FO 42-51683, (Squadron Unknown),

Arrived 28 August 1944, returned to the US on 10 June 1945.

B-24J-10-FO 42-51695, (767th Sq.)

Arrived 8 September 1944, crashed 13 February 1945.

42-51695

B-24J-10-FO 42-51759, (766th Sq., #41)

Arrived 6 September 1944, returned to the US on 11 June 1945.

B-24J-10-FO 42-51778, (765th Sq.)

Arrived 16 August 1944, transferred to the 885th Special Squadron on 30 August 1944.

B-24J-10-FO 42-51783, (765th Sq., #35)

What's Next? Arrived 7 September 1944, crash landed 28 April 1945.

B-24J-10-FO 42-51816, (764th Sq.)

Arrived 25 August 1944, transferred to 459th Bomb Group on 19 September 1944. Crashed on 4 June 1945.

B -24J-15-FO 42-51838, (767th Sq., #65)

42-51838

Natural, Arrived 7 September 1944, crashed due to weather on 31 March 1945. MACR 13485 (10 KIA).

B-24J-15-FO 42-51867, (765th Sq., #27)

42-51783

Arrived 27 August 1944, returned to US on 9 June 1945.

B-24J-15-FO 42-51894, (764th Sq.)

Arrived 11 October 1944, returned to the US on 11 July 1945.

B-24J-15-FO 42-51898, (767th Sq., #63)

Arrived 26 August 1944, shot down by flak over Blechhammer on 18 December 1944. MACR 10639 (10 POW).

B-24J-15-FO 42-51918, (766th Sq.)

Arrived 17 September 1944, crashed due to mechanical failure on 26 March 1945. MACR 13198 (10 POW).

B-24J-15-FO 42-51922, (766th Sq., #41)

Arrived 21 August 1944, lost to fighters over Trieste, Italy on 16 December 1944. MACR 10494 (10 POW).

B-24J-15-FO 42-51925, (764th Sq., #14)

Arrived 31 August 1944, transferred to 484th Bomb Group. Lost with that group 20 November 1944.

B-24J-15-FO 42-51937, (764th Sq., #5)

Arrived 11 October 1944, returned to the US on 24 July 1945.

B-24J-15-FO 42-51946, (764th Sq., #1)

42-51946

Arrived 15 December 1944, returned to the US on 8 June 1945.

B-24J-15-FO 42-51949, (766th Sq., #41)

Arrived 27 August 1944, returned to the US 14 June 1945.

B-24J-15-FO 42-51967, (765th Sq., #34

42-51967

Arrived 2 September 1944, transferred to the 484th Bomb Group, lost with that group on 18 December 1944.

PB-24J-15-FO 42-51971, (765th Sq.)

Arrived 13 October 1944, disappeared somewhere in the Alps during bad weather on 15 November 1944 MACR 9885 (10 KIA).

B-24J-15-FO 42-52025, (765th Sq., #26)

42-52025

Arsenic and Lace arrived 11 October 1944, lost to fighters on Odertal mission 17 Dec 1944. MACR 10680 (5 KIA, 5 POW).

B-24J-15-FO 42-52052, (765th Sq., #27)

Alma-M-Amur, Arrived 16 December 1944, returned to the US on 2 July 1945.

B-24J-15-FO 42-52053, (766th Sq.)

Arrived 3 January 1945, transferred to another group. Returned to US 12 June 1945.

B-24J-15-FO 42-52056, (766th Sq., #60)

Arrived 17 September 1944, returned to the US on 12 June 1945.

PB-24J-15-FO 42-52060, (Squadron Unknown),

Arrived 12 October 1944, returned to the US on 10 June 1945.

PB-24J-15-FO 42-52065, (766th Sq.)

Arrived 5 October 1944, transferred to the 465th Bomb Group 23 October 1944. Returned to the US on 8 June 1945.

B-24H-15-FO 42-52325, (767th Sq., #60)

42-52325

Bushwacker, Arrived 3 February 1944. Crash landed at Torretta returning from its final mission on 15 Nov 44. The nickname was painted on just before the final mission.

B-24H-15-FO 42-52336, (766th Sq., #49)

42-52336

The spirit of Hollywood/Gloria Jean, arrived on 31 January 1944, destroyed 13 April 1944 when another plane damaged by flak, crashed into 42-52336. MACR 4649 (2 KIA, 8 MIA).

B-24H-15-FO 42-52361, (765th Sq., #22)

Arrived 19 February 1944, severely damaged by flak on 6 April 1944, the pilot was able to reach the coast before bailing out. Three crew members, who bailed out earlier could not be found. MACR 3698 (3 MIA).

B-24H-15-FO 42-52366, (766th Sq., #44)

Arrived 31 January 1922, on 13 April 1944, flak tore off part of the wing causing it to flip over and into another aircraft. MACR 3974 (10 KIA).

B-24H-15-FO 42-52368, (765th Sq., #34)

42-52368

Piecemaker, arrived 3 March 1944, battle damaged during 6 April 1944 mission to Zagreb, Yugoslavia, repaired and returned to service, shot down by German fighters during 25 July 1944 mission to Linz, Austria. MACR 7110 (3 POW, 7 KIA).

B-24H-15-FO 42-52371, (765th Sq., #37)

42-52371

Upstairs Maid, arrived 2 March 1944, severely damaged by fighters during Linz mission on 25 July 1944, repaired and returned during October 1944, returned to the US on 27 June 1945.

B-24H-15-FO 42-52378, (767th Sq., #74)

Arrived 2 February 1944, transferred to 451st Bomb Group on 28 February 1944.

B-24H-15-FO 42-52379, (764th Sq., #4)

Arrived 2 February 1944. An original group aircraft, while taxiing after a training flight at Oudna, the co-pilot accidentally retracted the landing gear causing damage to the ship. The damage was considered to be “beyond economical repair”.

B-24H-15-FO 42-52388, (767th Sq., #62

Our Baby, On the first combat mission for the group the pilots were trying to observe the results of the bomb run when this aircraft collided with another ship causing it to crash. MACR 4086 (1 POW, 9 KIA).

B-24H-15-FO 42-52389, (766th Sq., #58)

42-52389

Jug-Butt, arrived 31 January 1944. Returning from a training mission on 28 March 1944, the plane touched down on the runway and the main gear collapsed causing the ship to swerve off the runway and causing even more damage. Condemned on that date. There were no injuries.

B-24H-15-FO 42-52393, (764th Sq., #2)

42-52393

Ignatz, Arrived 1 February 1044, suffered major accident on 29 February 1944 when it flew into the ground when trying to get below the weather.

B-24H-15-FO 42-52395, (767th Sq., #61)

Little Jesus, arrived 2 March 1944, lost to flak over Belgrade on 16 April 1944. MACR 4022 (5 KIA, 5 POW).

B-24H-15-FO 42-52396, (767th Sq., #68)

Flak Finder, Original 461st aircraft arrived 2 February 1944, transferred to 449th Bomb Group 3 March 1944 and lost with that group 29 May 1944.

B-24H-15-FO 42-52398, (766th Sq., #41)

42-52398

Boise Belle, Original aircraft arrived 2 February 1944, salvaged (combat damage) on 24 March 1944.

B-24H-15-FO 42-52399, (767th Sq., #73)

42-52399

Miss Carriage, Arrived 2 February 1944, crashed 27 May 1844 due to flak damage. MACR 5416 (4 POW, 6 Evadees).

B-24H-15-FO 42-52405, (766th Sq., #54)

Arrived 7 February 1944, lost to flak 24 May 1944. MACR 5041 (4 KIA, 6 POW).

B-24H-15-FO 42-52407, (767th Sq., #65)

42-52407

Arrived 19 February, salvaged (battle damage) 2 March 1945.

B-24H-15-FO 42-52408, (766th Sq., #43)

42-52408

The Kissed Off Kids Arrived 31 January 1944, crashed short of the runway returning from a practice mission on 24 August 1944. The wings were torn off and the fuselage was completely consumed by fire, but the crew was able to escape the aircraft. One crew member was seriously injured. Contrary to some rumors, it was never rebuilt.

B-24H-15-FO 42-52409, (764th Sq., #16)

Hell's A Poppin', Arrived 7 February 1944, lost to flak over Budapest on 13 April 1944. MACR 3873 (10 POW).

B-24H-15-FO 42-52412, (767th Sq., #75)

42-52412

Purple Shaft, arrived 2 March 1944, returned to the US on 9 June 1945.

B-24H-15-FO 42-52436, (766th Sq., #48)

42-52436

Down N' Go, Arrived 7 February 1944, Crashed due to mechanical problems on 7 August 1944. MACR 11267 (10 POWs).

B-24H-15-FO 42-52451, (766th Sq., #66)

Arrived 2 February 1944, shot down by fighters on 25 July 1944. MACR 7004 (6 POW, 4 KIA).

B-24H-15-FO 42-52458, (764th Sq., #13)

42-52458

Chippiedall, Arrived 20 February 1944, damaged by fire from enemy aircraft on 11 June 1944, #13 tried to reach the Isle of Vis. However, the crew was forced to bail out 30 miles south of Nis, Yugoslavia. All 11 men were able to evade capture and returned to base. MACR 6307.

B-24H-15-FO 42-52459, (764th Sq., #8)

Winona Belle, Arrived 19 February 1944, shot down by German fighters during 25 July 1944 mission to Linz, Austria. MACR 11978 (5 POW, 5 KIA).

B-24H-15-FO 42-52460, (764th Sq., #17

Red Ryder, Arrived 8 February 1944, transferred to the 451st Bomb Group.

B-24H-15-FO 42-52486, (765th Sq., #26)

42-52486

Irish Angel, Arrived 19 February 1944, severely damaged by German fighters during the 25 July 1944 mission to Linz, Austria, Lt, Wester was forced to bail out his crew over the friendly Isle of Vis.

B-24H-15-FO 42-52493, (765th Sq., #35)

Grimms Gremlins, Arrived 19 February 1944, crash landed 20 April 1944 as a result of blown nose tire.

B-24H-15-FO 42-52550, (765th Sq., #31)

42-52550

Arrived 13 February 1944, transferred to 47th Bomb Wing on 28 February 1944.

B-24H-15-FO 42-52723, (766th Sq., #58)

Transferred to 461st from 484th. Shot down by fighters on 12 July 1994. MACR 6894 (8 POW, 2 KIA).

B-24G-5-NT 42-78103, (764th Sq., #17)

Red Ryder, Arrived 16 April 1944, damaged by flak on 31 May 1944 Lt. Ryder tried to get to the Isle of Vis, but were forced to bail out six miles south of the island. All ten men drowned. MACR 5841.

B-24G-5-NT 42-78123, (764th Sq., #19)

42-78123

Fertile Myrtle, Arrived 1 April 1944, transferred to the 485th Bomb Group.

B-24G-10-NT 42-78176, (Squadron Unknown)

Arrived 19 April 1944, transferred to 451st 17 December 1944, returned to US with that group.

42-78176

B-24G-10-NT 42-78179, (766th Sq., #58)

Three Bad Wolves, Arrived 9 April 1944, crashed 24 April 1944.

42-78179

B-24G-10-NT 42-78202, (766th Sq., #50)

Hi-Ho Silver. Arrived 4 April 1944, Shot down by fighters 12 July 1944. MACR 6895 (4 KIA, 6 POW).

42-78202

B-24G-10-NT 42-78212, (764th Sq.)

Scrounch, arrived 4 March 1944, shot down by German aircraft over Wiener Neustadt on 10 May 1944, MACR 4657 (8 POW, 2 KIA).

B-24G-10-NT 42-78224, (767th Sq., #74)

42-78224

Arrived 18 April 1944, transferred to 484th Bomb Group.

B-24G-10-NT 42-78228, (765th Sq., #31)

The Iron Duke, Arrived 17 April 1944, developed engine trouble and crashed into the sea 22 May 1944. The entire crew perished. MACR 4917.

B-24G-10-NT 42-78247, (765th Sq., #39)

42-78247

Arrived 27 April 1944, disabled by flak on the 4 October 1944 mission to Munich, the aircraft made an emergency landing at Lansberg-Penzing Field. The ship was repaired and used by German forces for experimental flights. MACR 8972.

B-24G-10-NT 42-78253, (765th Sq., #29)

Arrived 23 April 1944, crashed 22 July 1944.

B-24G-10-NT 42-78260, (767th Sq., #71)

Arrived 24 April 1944, Hit by flak on 10 June 1944 over Guirgiu, Romania. MACR 5641 (10 POW).

B-24G-10-NT 42-78267, (767th Sq., #72)

42-78267

Wolf Den, Arrived 1 May 1944, engine trouble due to flak over Weiner Neustadt on 24 May 1944. MACR 5056 (10 POW).

B-24G-10-NT 42-78286, (764th Sq. #10)

42-78286

Arrived 2 May 1944, crashed 16 March 1945.

B-24G-10-NT 42-78291, (766th Sq., #51)

Arrived 1 May 1944, shot down by fighters 10 June 1944. MACR 7034 (10 POW).

B-24G-15-NT 42-78332, (766th Sq., #54)

42-78332

Arrived 27 May 1944, ran out of fuel on mission to Bucharest 26 August 1944, Bailed out crew along the coast. MACR 7962 (10 POW).

B-24G-16-NT 42-78357, (766th Sq.)

Arrived 21 May 1944, Crashed due to mechanical failure 15 May 1945.

B-24G-16-NT 42-78408, (766th Sq., #49)

42-78408

Arrived 1 July 1944, Damaged by flak over Vienna on 21 February 1945. Lt. Hooper attempted to reach Russian lines, but was forced to bail out the crew over Hungary when they ran out of fuel. They were interviewed by three Hungarian officers and turned over to the commander of the SS Totenkopf (Death's Head) Division. They were never seen again. MACR 12359 Note: serial number on MACR (42-58408) is incorrect.

B-24G-16-NT 42-78417, (766th Sq., #44)

42-78417

Bottoms Up, Transfer from 376th Bomb Group (war weary), crashed 30 April 1945.

B-24G-16-NT 42-78436, (766th Sq., #52)

42-78436

Shady Lady, Arrived 10 July 1944, transferred to 451st Bomb Group.

B-24G-16-NT 42-78444, (765th Sq., #24)

42-78444

All American, Transfer from 485th Bomb Group, crashed into the Adriatic attempting to get home. MACR 8973 (I KIA, 9 POW).

B-24G-16-NT 42-78446, (766th Sq., #47)

42-78446

Urgent Virgin, Transfer from 485th Bomb Group, lost to flak on 4 October 1944.

B-24G-16-NT 42-78466, (766th Sq.)

Arrived 6 July 1944. Lost to flak on 4 October 1944.

B-24J-1-NT 42-78497, (764th Sq., #6)

42-78497

Bubble Trouble, Arrived 3 August 1944, transferred to the 451st Bomb Group on 21 November 1944.

B-24J-1-NT 42-78499, (767th Sq.)

Arrived 17 July 1944, crashed 16 February 1945.

B-24J-1-NT 42-78519, (767th Sq., #75)

Arrived 31 July 1944, damaged by flak 17 August 1944, ditched in the Adriatic attempting to return to Torretta. All crew members were picked up by German E boats. MACR 8260 (10 POW).

B-24J-1-NT 42-78591, (766th Sq., #48)

Arrived 31 July 1944, mid-air collision 13 October 1944.

B-24J-1-NT 42-78606, (765th Sq., #27)

Betty Jean, Transfer from 451st on 21 October 1944, lost 26 December 1944.

42-78606

B-24J-1-NT 42-78616, (765th Sq., #29

Shack Wagon, Arrived 4 August 1944, hit by flak 20 January 1945, crashed landing, salvaged.

42-78616

B-24J-1-NT 42-78627, (765th Sq., #20)

Arrived 12 September 1944, mid-air collision 13 October 1944.

42-78627

B-24J-1-NT 42-78676, (764th Sq., #2)

Arrived 16 September, mid-air collision 9 March 1945.

42-78676

B-24J-1-NT 42-78677, (764th Sq., #17)

42-78677

Arrived 16 September 1944, returned to the US 8 June 1945.

B-24H-15-FO 42-94729, (764th Sq., #2)

42-94729

The One-Eyed Jack, Transfer from 484th Bomb Group, crash landed short of the runway 23 December 1944 (ferry mission).

B-24H-15-FO 42-94732, (767th Sq., #68)

42-94732

Tail Dragon, Transfer from 484th Bomb Group, shot down by fighters 25 July 1944 MACR 7037 (3POW, 7KIA).

B-24H-25-FO 42-95257, (767th Sq., #72)

Arrived 10 May 1944, shot down by fighters 25 July 1944 MACR 7117 (10 POW).

B-24H-25-FO 42-95287, (766th Sq., #57)

Strictly G.I., Arrived 10 May 1944, shot down 19 November 1944 MACR 9937 (10 POW).

B-24H-30-FO 42-95304, (766th Sq., #42)

42-95304

Arrived 10 May 1944, shot down by enemy aircraft 17 December 1944 MACR 10701 (10 POW).

PB-24H-30-FO 42-95344, (767th Sq., #70)

Arrived 6 July 1944, right main landing gear and nose wheel collapsed upon landing 30 May 1945.

B-24H-30-FO 42-95359, (Squadron Unknown)

42-95359

Transferred from 451st Bomb Group, salvaged 21 January 1945.

B-24H-30-FO 42-95383, (764th Sq., #12)

42-95383

Lazy Lady, Arrived 1 June 1944, shot down by fighters 25 July 1944 MACR 7035.

B-24J-15-CF 42-99759, (766th/764th Sq., #56/11)

42-99759

Cherokee, Transfer from 466th BG, returned to the US on 8 June 1945.

B-24J-60-CF 44-10550, (764th Sq., #5).

44-10550

Miss Kay, Transfer from 484th Bomb Group on 25 November 1944, returned to the US on 25 June 1945.

B-24J-60-CF 44-10557, (765th Sq., #27)

44-10557

You Bet!, Transfer from 8th Air Force, returned to the US on 15 June 1945.

B-24J-70-CF 44-10628, (Squadron Unknown)

Arrived 12 August 1944, returned to the US on 9 June 1945.

B-24J-155-CO 44-40269, (766th Sq., #55)

Rough Rid'er, Transfer from the 484th Bomb Group, Returned to the US on 8 June 1945.

44-40269

B-24J-170-CO 44-40592, (767th Sq., #71)

The Mary E, arrived 30 May 1944, landing accident destroyed the aircraft on 30 August 1944.

44-40592

B-24J-170-CO 44-40621, (764th/767th Sq., #6)

Arrived 30 May 1944, blew up over target on 30 August 1944.

B-24J-170-CO 44-40632, (764th Sq.)

Arrived 6 June 1944, severely damaged by flak over the Herman Goering Tank Works with Lt. Douglas Herrin at the controls, it was condemned after the aircraft crash landed at Torretta Field.

B-24J-185-CO 44-40896, (764th Sq., #4)

Arrived 30 June 1944, direct hit blew plane to bits 4 October 1944.

44-40896

B-24J-190-CO 44-41016, (765th Sq. #33)

44-41016

Arrived 9 August 1944, shot down by enemy aircraft 17 December 1944 MACR 10636 (4 POW 6 KIA).

B-24J-190-CO 44-41020, (765th Sq., #23)

44-41020

Judy R, Arrived 17 July 1944.

B-24J-190-CO 44-41039, (764th Sq., #1)

44-41039

Male Box, Arrived 17 July 1944, shot down by enemy aircraft MACR 8976 (10 POW).

B-24J-190-CO 44-41044, (764th Sq., #18)

44-41044

Lady Duzz, Arrived 19 July 1944.

B-24J-190-CO 44-41046, (764th Sq.)

Arrived 17 July 1944.

B-24J-195-CO 44-41069, (767th Sq., #72)

Ford's Folly, Arrived 6 August 1944, hit by flak 25 September 1944.

B-24J-195-CO 44-41091, (764th Sq., #7)

Sweet Adeline, Arrived 31 July 1944, hit by flak 23 March 1944 (10 POW).

B-24J-195-CO 44-41093, (766th Sq., #45)

Arrived 28 July 1944, hit by flak 16 October 1944 MACR 9192.

B-24J-195-CO 44-41112, (765th Sq., #26)

44-41112

Irish Angel II/She Wolf, Arrived 4 August 1944, returned to the US on 8 June 1945.

B-24J-195-CO 44-41113 (765th Sq., #22)

Arrived 22 July 1944, lost to fire 11 September 1944.

44-41113

B-24J-195-CO 44-41120, (767th Sq., #61)

Arrived 1 July 1944, transferred to 484th Bomb Group.

44-41120

B-24J-195-CO 44-41140, (767th Sq., #76)

Arrived 12 August 1944, take off accident 15 November 1944.

44-41140

B-24J-200-CO 44-41150, (squadron Unknown)

Arrived 13 August 1944, salvaged 6 November 1944.

B-24J-200-CO 44-41156, (766th Sq., #51)

Arrived 4 August 1944, take off accident 5 September 1944.

44-41156

B-24J-200-CO 44-41158, (767th Sq., #77)

Arrived 8 August 1944, hit by flak 17 December 1944 MACR 10650.

B-24J-200-CO 44-41162, (765th Sq., #38)

Stumpy Joe, Arrived 21 July 1944, gas leak in wing tank 11 November 1944 MACR 10490.

44-41162

B-24J-200-CO 44-41212, (Squadron Unknown)

Arrived 20 August 1944, returned to the US on 7 June 1945.

B-24J-205-CO 44-41338, (766th Sq., #56)

44-41338

Cherokee, Arrived 17 August 1944, crash landing 11 December 1944.

B-24J-205-CO 44-41355, (764th Sq., #14)
Arrived 17 September 1944, return to the US on 6 June 1945.

B-24J-20-FO 44-48757, (767th Sq., #71)
Arrived 17 September 1944, mechanical problems 8 January 1945.

B-24J-20-FO 44-48761, (766th Sq., #52)
Arrived 29 November 1944, hit by flak 20 January 1945 MACR 11805 (2 POW, 8 KIA).

PB-24J-20-FO 44-48796, (Squadron Unknown)
Arrived 12 October 1944. Returned to the US on 16 June 1945.

PB-24J-20-FO 44-48833, (765th Sq.)
Arrived 12 October 1944, Returned to the US on 8 June 1945.

B-24J-20-FO 44-48993, Gene's (765th Sq., #24)

44-48993

Hairpower, Arrived 29 November 1944. Shot down by fighters over Vienna on 21 February 1945. MACR 12475 (10 POW).

PB-24L-1-FO 44-49038, (765th Sq., #30)

Billie K, Arrived 12 October 1944, Returned to the US on 8 June 1945.

Image9924.JPG

PB-24L-1-FO 44-49087, (Squadron Unknown)

Arrived 11 December 1944, Returned to the US on 28 June 1945.

PB-24L-1-FO 44-49242, (764th Sq., #19)

44-49242

Arrived 31 October 1944. Returned to US on 31 May 1945. It was the favorite ship of Capt. Ernie Parsonson, which he also flew home after the war. Landed at Hunter Field, GA., was declared excess the next day, and was scrapped at Walnut Ridge, Ark.

B-24L-5-FO 44-49346, (764th Sq., #12)

Arrived 31 January 1945, Returned to the US 9 June 1945.

44-49346

B-24L-5-FO 44-49375, (765th Sq., #32)

44-49375

Arrived 4 December 1944, Returned to the US on 6 June 1845.

B-24L-5-FO 44-49390, (767th Sq., #79)

Arrived 4 December 1944, Returned to the US on 12 June 1945.

B-24L-5-FO 44-49395, (Squadron Unknown)

Arrived 7 February 1945, Returned to the US on 17 June 1945.

B-24L-5-FO 44-49407, (Squadron Unknown)

Arrived 5 December 1944, Returned to US on 8 June 1945.

B-24L-5-FO 44-49413, (Squadron Unknown)

Arrived 30 December 1944, Returned to the US.

B-24L-5-FO 44-49428, (765th Sq., #29)

Arrived 1 February 1945, Crashed due to engine trouble 26 March 1945.

B-24L-5-FO 44-49501, (767th Sq.)

Red Head, Arrived 19 December 1944, Ditched in Adriatic Sea, out of fuel. MACR 14228 (4 Rescued, 8 KIA).

B-24L-10-FO 44-49511, (764th Sq., #15)

44-49511

Miss Lace, Arrived 7 February 1945, Lost to flak over Linz on 25 April 1945. Nine of the survivors were wounded in the attack. This was the last aircraft lost by the group. MACR 15048 (1 KIA, 10 POW).

B-24L-10-FO 44-49534, (Squadron Unknown)

Arrived 7 February 1945, Returned to the US on 9 August 1945.

B-24L-10-FO 44-49541, (Squadron Unknown)

Arrived 1 February 1945, Returned to the US on 8 June 1945.

B-24L-10-FO 44-49572, (Squadron Unknown)

Arrived 2 January 1945, Returned to US on 29 July 1945, Returned to the US on 29 July 1945.

B-24L-10-FO 44-49584, (767th Sq., #71)

Arrived 16 December 1944, Taxi accident, slid off muddy runway on 31 January 1945.

B-24L-10-FO 44-49586, (Squadron Unknown)

Arrived 8 December 1944. Returned to the US on 8 June 1945.

B-24L-10-FO 44-49588, (Squadron Unknown)

Arrived 15 December 1944. Returned to the US on 8 June 1945.

B-24L-10-FO 44-49619, (Squadron Unknown)

Arrived 3 December 1944. Return to US on 16 June 1945.

B-24L-10-FO 44-49623, (767th Sq., #78)

Arrived 5 December 1944, Transferred to the 451st Bomb Group.

B-24L-10-FO 44-49641, (767th Sq., #63)

Arrived 5 December 1944, Mechanical trouble over Strasshof, Austria on 26 March 1945, MACR 13192 (10 POW).

B-24L-10-FO 44-49645, (767th Sq., #77)

44-49645

Arrived 15 December 1944, Returned to the US on 7 June 1945.

B-24L-10-FO 44-49674, (765th Sq.)

Maggie's Boomerang, Arrived 2 January 1945, Return to US on 12 June 1945.

B-24L-10-FO 44-49679, (764th Sq.)

Arrived 18 December 1944, Transferred out on 31 January 1945.

B-24L-15-FO 44-49764, (Squadron Unknown)

Arrived 23 December 1944, Returned to the US on 11 June 1945.

PB-24L-15-FO 44-49851, (Squadron Unknown)

Arrived 19 February 1945, Returned to the US on 18 June 1945.

B-24L-15-FO 44-49879, (Squadron Unknown)

Arrived 19 January 1945, Returned to the US on 18 June 1945.

B-24L-15-FO 44-49887, (764th Sq., #6)

Bingo, Arrived 19 January 1945, Transferred to the 465th Bomb Group.

44-49887

B-24L-15-FO 44-49896, (765th Sq., #23)

Rhode Island Red II, Arrived 19 January 1945. Returned to the US on 2 June 1945.

44-49912

PB-24L-15-FO 44-49921, (Squadron Unknown)
Arrived 19 January 1945. Returned to the US on 24 June 1945.

B-24L-15-FO 44-49925, (766th Sq., #52)
Arrived 19 January 1945. Hit by flak over Vienna on 13 February 1945 MACR 12489 (10 POW).

B-24L-15-FO 44-49952, (766th Sq., #75)
Arrived 7 February 1945. Returned to the US on 9 June 1945.

B-24L-15-FO 44-49984, (Squadron Unknown) Arrived 20 January 1945. Returned to the US on 7 June 1945.

B-24M-1-FO 44-50292, (Squadron Unknown)
Arrived 17 March 1945. Returned to the US on 9 June 1945.

B-24M-1-FO 44-50386, (767th Sq.)
Arrived 18 February 1945. Returned to the US on 10 June 1945.

B-24M-1-FO 44-50394, (Squadron Unknown)
Arrived 14 March 1945. Returned to the US on 8 June 1945.

B-24M-1-FO 44-50410, (764th Sq., #5)
Bingo Arrived 14 February 1945, mid-air collision 9 March 1945 MACR 12945.

B-24M-1-FO 44-50428, (766th Sq.)

44-50428

Arrived 12 February 1945. Returned to the US on 8 June 1945.

PB-24M-5-FO 44-50616, (765th Sq., #39)

44-50616

Arrived 19 February 1945. Returned to the US on 8 June 1945.

PB-24M-10-FO 44-50712, (Squadron Unknown)
Arrived 2 March 1945. Returned to the US on 7 June 1945.

B-24M-10-FO 44-50735, (765th Sq., #33)

44-50735

Arrived 7 March 1945. Returned to the US on 6 June 1945.

B-24M-10-FO 44-50743, (765th Sq., #28)
Arrived 7 March 1945. Shot down by enemy aircraft on 11 April 1945.

B-24M-10-FO 44-50814, (Squadron Unknown)
Arrived 12 March 1945. Returned to the US on 9 June 1945.

{4}

MISSIONS

MISSION PLANNING

MISSION SUMMARIES

THE HARD MISSIONS

SUPPLY MISSIONS

MISSION PLANNING

INTELLIGENCE

THE SUCCESS of daily bombing missions was dependent upon the latest information about the target, enemy flak defenses, and the abilities of the German Air Force (GAF). For the Fifteenth Air Force, given the number of countries within bombing range, and the variety of Allied military organizations contributing information, a thorough analysis of data became somewhat difficult. In fact, the S-2 Section (Intelligence) at Fifteenth Air Force Headquarters had a distribution list of no less than 65 cooperating intelligence agencies in the Mediterranean alone.

Intelligence gathering did not begin with the creation of the Fifteenth Air Force. The Northwest African Strategic Air Force had flown against Tunisia, Sardinia, Italy, and southern Europe. The 98th and 376th Bomb Groups, then with the 9th Air Force had attacked Ploesti, Libya, and Italy. The Western Desert Air Force supporting the British Eighth Army, and the Northwest African Tactical Air Force had been fighting the German and Italian Air Forces in Tunisia and Libya. Both the staff and combat elements of these units, for a full year, collected intelligence and learned how to use that information in planning and carrying out air operations.

They had also benefitted from interrogating enemy prisoners, briefing air crews, examining crashed and captured enemy aircraft, briefing air crews how to evade capture should they be shot down, compiling better maps. The year had been spent moving from a fledgling to an experienced staff upon which the Fifteenth could depend.

The Fifteenth Air Force S-2 Section, as well as the intelligence sections at wing and group headquarters were responsible or five areas: collection, analysis, and presentation of information about the enemy, the collection of operational intelligence materials, reporting of operations, counter-intelligence, and repatriation. Every morning the Commanding General and Deputy Commanding General were briefed on new developments in the GAF. A presentation was made to the Headquarters staff on the air, ground, and naval situation. The intelligence staffs at wing and group were also informed through appropriate channels. However, the primary activity was the analysis of information on strategic target systems such as airfields, oil refineries, or marshalling yards. No target was recommended by S-2 until the data had been thoroughly analyzed and put into a presentable format.

The Intelligence Section of Air Staff (A-2), passed this intelligence to wing commanders for study by their staffs. Intelligence officers at wing level were responsible for keeping the commander informed, providing whatever information was necessary for staff planning, and for providing assistance at the group level. Finally, the groups would put the intelligence to practical use. The most important responsibility of group intelligence officers was the daily briefing for pilots and air crews on the information necessary to complete their assignments.

At this point all the intelligence officers could do was wait for the aircrews to return and learn the accuracy of all their efforts. The first results came from a telephonic "Flash Report" based on interrogations of the first few crews to return. It was only a preliminary report and not forwarded to higher command. A second, more detailed phone report, based on the interrogation of all crews would be sent to Wing Headquarters within a few hours of the last planes return. Finally, a written narrative (Daily Operations Report, Form 34) was prepared by each group and sent to Headquarters for review by operations, statistical, and other staff offices. Based on all the information received, the Intelligence and Operations Summary (Intops Summary) was produced daily and sent to all subordinate commands along with a long list of other organizations.

OPERATIONAL PLANNING

As was mentioned earlier, the strategic goals of the Fifteenth Air Force was crippling or destroying German aircraft fighter strength, destroying petroleum, oil, and lubrication (POL) facilities, destroying railyards and other transportation targets, and support for ground forces. The vast majority of these targets were large, permanent, unchanging in function, and numerous. Therefore, long range planning was possible according to the type of target. For example, the destruction of the German Air Force (GAF) could take the form of attacks on fighter aircraft, airdromes, aircraft or aircraft engine manufacturing plants, or aviation fuel. An assault on the oil industry would involve oil fields and refineries, oil storage facilities, or rail yards and rolling stock (oil cars). The latter would also address attacks on transportation capabilities. The decision what type of target to attack came down to the Fifteenth in general directives from the Combined Chiefs of Staff. The authority for these directives came from their deputy for air operations, the Commanding General, USSTAF (United States Strategic Air Forces).

FIFTEENTH AIR FORCE OPERATIONS

Division of Combat Effort by Type Target

1 November 1943 – 8 May 1945

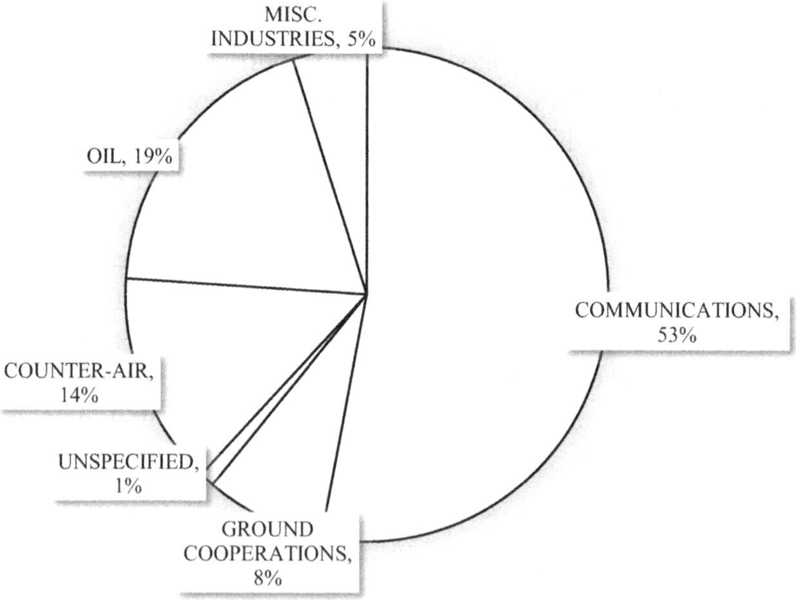

Percentages are based on tons of bombs dropped and have been taken from the statistical story of the Fifteenth Air Force, 28 SCU, May 1945

Actual operational planning took place at Headquarters, Fifteenth Air Force following daily reviews of the most recent information on the target system chosen, and a tactical decision arrived at after careful consideration of factors involved in planning an immediate operation. For the most part, operations involved between 500 and 900 bombers escorted by 300-400 fighters. The time to plan such an operation would run from twelve to eighteen hours. Once the mission commenced there would be no opportunity for second guesses and last-minute changes.

BOMB LOADS

There were a number of factors that could bring about changes to the mission. Bombing accuracy had an extreme effect on the number of bombers necessary to destroy the target. Bombing accuracy and target selection would also determine the type of ordnance loaded on the aircraft. Fragmentation bombs were used against parked aircraft, and more rarely, against gun emplacements and troop concentrations, where the bomb pattern needed to be spread over a large area. Groups arriving later were also more able to pick up the target due to the relatively small amount of smoke caused by frags. Heavy fragmentation bombs with proximity fuses were also used with success when deployed over flak batteries.

Incendiary bombs were employed primarily against oil refineries, rolling stock (oil cars), and industries considered highly flammable. Like fragmentation bombs it was spread over a large area, but caused a huge amount of smoke, making it impossible for following groups to see the target.

The greatest percentage of missions were flown with high explosives (HE). There were actually several types of high explosive bombs and planners had to take into consideration the characteristics of the target to maximize the effects of the selected bombload. As with other types of ordnance, smaller bombs could cover a larger area while larger bombs would be used to attack individual targets.

WEATHER RECONNAISSANCE

Weather was followed closely and could result in changes to the mission. Every day weather briefings were held with the Fifteenth Air Force Operations Officer. A morning briefing was held to discuss the probable weather the following day. An eleven o'clock briefing provided a detailed forecast at the planning meeting. There was also a short midnight briefing. There was a final briefing just before takeoff based on reports from reconnaissance aircraft already airborne. Based on this report, the Operations Officer, along with the Commanding General, issued an order to commence the mission, or the order to stand down.

BOMBER TACTICS

Generally, the Fifteenth Air Force organized missions one of two ways. Closely timed mass operations were flown against heavily defended targets, whereas loosely-timed attacks were employed against minor targets which were assumed to be lightly defended. The deciding factor was the anticipated level of enemy resistance.

On the morning of the mission the group leader took off first and circled the airfield gaining altitude until he reached the formation altitude of between 2,000 and 10,000 feet, depending on weather. Each aircraft took off at 30-second intervals and joined up with the group leader until all the ships were in their assigned position in the formation, a procedure that took between 30 and 60 minutes. A usual formation consisted of four boxes of six aircraft per box. The wing assembled next, made up of combat waves consisting of three to five groups. Each wing would then leave its assigned assembly point at a specific time to assure they arrived at the Control Point at the correct time to fall in line behind the previous wing. This single column was the usual formation when attacking the same target. The normal interval between groups was two minutes. This meant a one-group column was 6 or 7 miles long. If all 21 bomb groups of the Fifteenth Air Force were in the column, it would be 125 – 145 miles long. However, there were certainly a number of variations to the single column:

- Changes in climb and assembly might be caused by weather at takeoff or distance to target,
- To save gas a low altitude assembly would be made with a gradual climb to the target,
- A high-altitude assembly would be more common for targets that were closer to their home base,
- There may be 2, 3, or 4 columns depending on the number of targets being attacked,
- Bombing altitude changed according to enemy defenses and weather conditions,
- The length and direction of the route out and back were altered by flak defenses, wind direction and velocity, enemy fighter opposition, the position of the sun, smoke screens, and other factors.

After October 1944, a plan was devised to divide the group into two attack forces. This was usually a result of an insufficient number of fighters to cover the entire formation, or not enough PFF aircraft to support a maximum strength effort. To address this problem a normal strength force with four PFF ships per group and sufficient fighter coverage, designated Red Force, was sent against a major target. By contrast, Blue Force was sent out without fighter escort, and would have been assigned a less well defended target where visual bombing techniques could be used.

MISSION SUMMARIES

April 1944

THE GROUP would fly sixteen missions, putting up 508 sorties during April 1944. Targets attacked were in Yugoslavia, Italy, Hungary, Romania, Austria, and France. They dropped a total of 874 tons of bombs. Gunners claimed six enemy aircraft shot down, with five more probably destroyed and another three damaged

On April 2, 1944, the 461st flew its first combat mission when it struck the marshalling yards at Bihac, Yugoslavia. The weather was clear and the results goods were good. Unfortunately, the planes flown by 1Lt William Zumsteg and 2Lt Sidney Wilson were involved in a mid-air collision while watching the results of the bomb run. As a result, two ships and crews were lost.

<table>
<tr><th>1 Bihac, YUG - Marshalling Yard 2/4/1944</th><th>ESCORT</th><th>WEATHER</th><th>LOSSES</th></tr>
<tr><td rowspan="4">The 461st, on its freshman mission, dropped 61.5 tons of 20lb frags from 17,000/20,000 feet. 5 A/C jettisoned 9.5 tons, while 1 A/C dropped 1 ton on PIANOSA ISLAND, Six additional A/C returned early. Four E/A were seen in the target area, but there were no encounters. Two B-24s were lost due to collision at the rally point. Results reported by crews indicate 40 - 60 bombs in target area with heavy smoke in South part of marshalling yard.</td><td>None</td><td>CAVU</td><td>2-0-0</td></tr>
<tr><td>BOMBS</td><td>BOMBING</td><td>VICTORIES</td></tr>
<tr><td>20lb Frag</td><td>Visual</td><td>0-0-0</td></tr>
<tr><td colspan="2">FUSES
.1 Nose</td><td>FLAK
None</td></tr>
</table>

2 Drnis, YUG - Marshalling Yard 3/4/1944

27 A/C dropped 54 tons of 500lb bombs, 2 A/C dropped 4 tons on PEROVIC. Three A/C returned early. No E/A were seen, no encounters, no flak, no losses. Photos show 8 bursts 2000 feet East of choke point, the remaining bombs dropped SE of target and 3,000 feet SE of choke Point.

ESCORT	WEATHER	LOSSES
None	8/10	0-0-0
BOMBS	BOMBING	
500lb GP		0-0-0
FUSES		FLAK
.1 Nose, .025 Tail		None

3 Nis, YUG - Marshalling Yard 5/4/1944

36 A/C dispatched to attack the NIS M/Y were unable to locate the target due to weather. 34 A/C dropped 67 tons of 500lb bombs from 20,000 feet on this target. Two A/C returned early. Bomb strike photos show hits on RR and road to LEKOVIC with considerable damage to both.

ESCORT	WEATHER	LOSSES
None	8/10	0-0-0
BOMBS	BOMBING	VICTORIES
500lb GP		0-0-0
FUSES		FLAK
.1 Nose		I-A-H

4 Zagreb, YUG - Airdrome 6/4/1944

58 B-24s of the 460th and 461st Bomb Groups were dispatched to attack this target. Due to heavy overcast only 19 A/C dropped 3180 x 2lb frags from 20,500 - 22,000 feet. 39 A/C either jettisoned their bombs or returned them to their base. The 461st claimed one E/A destroyed.

ESCORT	WEATHER	LOSSES
1, 14, 82FG	9/10	0-0-0
BOMBS	BOMBING	VICTORIES
120lb frags	Pathfinder	1-0-0
FUSES		FLAK
.1 Nose		None

5 Ferrera, ITA - Marshalling Yard 7/4/1944

All 33 A/C of the 461st bombed the primary dropping 66 tons of 500lb GP bombs at 21,000. Photos show bombs well concentrated on target area. Track cut, hits on locos, rolling stock, RR bridge and track entrance to bridge, ball bearing plant, sugar refinery and flour mill.

ESCORT	WEATHER	LOSSES
1, 14, 82 FGs	CAVU	0-0-0
BOMBS	BOMBING	VICTORIES
500lb GP	Visual	0-0-0
FUSES		FLAK
.1 Nose, .025 Tail		I-A-H

6 Zagreb, YUG - Airdrome/Marshalling Yard 12/4/1944	ESCORT None	WEATHER 4/10	LOSSES 0-0-0
Hits on M/Y, assembly shops and three direct hits on E end of chock point in the yard. Control tower believed destroyed. Two assembly shops in SE corner of A/D destroyed. Two A/C believed destroyed. String of bombs fell across dispersal area. Admin building and sub-assembly buildings received hits.	BOMBS 20lb frags	BOMBING Pathfinder	VICTORIES 0-0-0
	FUSES .1 Nose, .025 Tail		FLAK S-I-H
7 Budapest, HUN - Duna Aircraft Factory 13/4/1944	ESCORT P-38s	WEATHER 4/10	LOSSES 3-3-0
Strike photos show the factory are to be well covered. Four assembly buildings to the West were well hit. A concentration of bombs fell SE of the target along river bank.	BOMBS 500lb GP	BOMBING Pathfinder	VICTORIES 3-3-0
	FUSES .1 Nose, .025 Tail		FLAK H-I-A
8 Bucharest ROU - Chitila Marshalling Yard 15/4/1944	ESCORT 14FG	WEATHER 9/10	LOSSES 0-0-0
Target completely obscured by cloud cover. Bombing done by instruments with unobserved results.	BOMBS 500lb GP	BOMBING Pathfinder	VICTORIES 0-0-0
	FUSES .1 Nose, .025 Tail		FLAK L-I-H
9 Belgrade, YUG - Zemun Airdrome 16/4/1944	ESCORT 306 FW	WEATHER 6/10	LOSSES 1-2-6
28 A/C dropped 5140 x 20lb fragmentation bombs fused instantaneous. Incomplete results indicate started on runway and carrying WSW through hangers into Rogasarski A/C factory.	BOMBS 20lb frag	BOMBING	VICTORIES 0-0-0
	FUSES Instantaneous		FLAK I-A-H
10 Belgrade, YUG - Zemun Airdrome 17/4/1944	ESCORT 325FG	WEATHER 10/10	LOSSES 0-0-2
460th and 461st encountered heavy overcast enroute to the target and returned to base.	BOMBS 20lb frag	BOMBING Did Not Bomb	VICTORIES 0-0-0
	FUSES Instantaneous		FLAK None

11 Tagliamento Casarsa, ITA - Bridges (Alternate) 20/4/1944			
Dropped 500lb GP bombs through the clouds from 13,000 - 21,500. Bomb strike photos show heavy cloud cover, however direct hits are seen on East end of both R.R. bridges. Hits are seen on East side of river. West side of river is completely obscured by clouds.	ESCORT 1FG, 52FG	WEATHER 7/10	LOSSES 0-2-1
	BOMBS 500lb GP	BOMBING	VICTORIES 0-0-0
	FUSES .1 Nose, .025 Tail		FLAK S-I-H

12 Bucharest, ROU - Chitila Marshalling Yard 21/4/1944			
Dropped 500lb GP bombs from 21,000 to 23,600. Bombing was effected by instruments through complete undercast. No results were observed and no photos were taken.	ESCORT 1FG, 52FG	WEATHER 10/10	LOSSES 0-0-1
	BOMBS 500lb GP	BOMBING Did Not Bomb	VICTORIES 1-0-0
	FUSES .1 Nose, .025 Tail		FLAK S-I-H

13 Bad Voslau, AUT - Airdrome 23/4/1944			
Dropped 500lb "Comp. B" .1 and .01 bombs and 12,550 x 20lb fragmentation bombs 19,000 - 23,000. A/D and all facilities heavily hit. Severely damaged buildings include all eight hangers, offices, barracks, boiler house and M/T bays.	ESCORT 306FW	WEATHER 3/10	LOSSES 1-4-10
	BOMBS Frags	BOMBING	VICTORIES 0-2-0
	FUSES Instantaneous		FLAK I-A-H

14 Bucharest, ROU - Chitila Marshalling Yard 24/4/1944			
Bomb strike photos show the round house was hit and numerous hits were scored along the W. choke point of the M/Y. Three strings extended across the RR tracks. Many strikes were scored in the main M/Y extending into the SE choke point of the small M/Y. Some bombs fell in built up area NE of the M/Y.	ESCORT P-38, 47, 51	WEATHER CAVU	LOSSES 1-0-0
	BOMBS 500lb GP	BOMBING Visual	VICTORIES 1-0-1
	FUSES .1 Nose, .025 Tail		FLAK I-A-H

<table>
<tr><td>15 Toulon Harbor, FRA - Submarine Pen
29/4/1944</td><td rowspan="2">ESCORT
P-38s,
P-51s</td><td rowspan="2">WEATHER
CAVU</td><td rowspan="2">LOSSES
0-0-6</td></tr>
<tr><td rowspan="3">An effective chemical smoke screen prevented visual observation of results and detailed interpretation of bomb strike photos is difficult. These show that M/Y just NW of LA SEYNE BAY sustained two direct hits as did the ammunition factory nearby. 70/80 hits scored SW of the oil storage facilities. Hits scored on the RR tracks and road intersection.</td></tr>
<tr><td>BOMBS
1000lb GP</td><td>BOMBING
Visual</td><td>VICTORIES
0-0-2</td></tr>
<tr><td colspan="2">FUSES
.1 Nose, .025 Tail</td><td>FLAK
I-A-M</td></tr>
<tr><td>Alessandria, ITA - Marshalling Yard 30/4/2944</td><td rowspan="2">ESCORT
1FG</td><td rowspan="2">WEATHER
CAVU</td><td rowspan="2">LOSSES
0-0-0</td></tr>
<tr><td rowspan="3">16 Bomb strike photos show strikes in vicinity of car repair shops, RR tracks and nearby buildings. The West end of the M/Y near choke point was hard hit. Also hits to the marshalling yard, car shop and munitions plant.</td></tr>
<tr><td>BOMBS
500lb GP</td><td>BOMBING
Visual</td><td>VICTORIES
0-0-0</td></tr>
<tr><td colspan="2">FUSES
.1 Nose, .025 Tail</td><td>FLAK
None</td></tr>
</table>

Eight group aircraft were lost during April 1944, six shot down by flak batteries and two non-combat losses due to the mid-air collision. A nineth ship had been reported lost, but was found to have safely landed on the friendly Isle of Vis.

Casualties included 31 men killed in action, 10 wounded in action, and another 40 who became prisoners of war.

May 1944

The group flew 742 sorties over nineteen missions, with early returns totaling 67 aircraft. They dropped 1,477 tons of bombs on a variety of targets, including marshalling yards, airdromes, harbors, and oil refineries. Seventeen enemy aircraft were claimed destroyed with another ten probable. Two others were reported damaged.

17 LaSpezia, ITA - Harbor Installations 2/5/1944	ESCORT	WEATHER	LOSSES
461st photos not received.	48 P-47s	UNDERCAST	0-0-5
	BOMBS	BOMBING	VICTORIES
	500lb GP	Visual	0-0-0
	FUSES		FLAK
	.1 Nose, .025 Tail		M-A-H
18 Ploesti, ROU - Marshalling Yard 5/5/1944	ESCORT	WEATHER	LOSSES
Many direct hits on tracks and rolling stock at RR junction South of marshalling yard. Direct hits in M/Y west of Concordia Vega Refinery, on the Ploesti North Station and in the SE corner of the Boosting Station area of Sterna Romana Refinery. Bombs carried into the Astro/Romano and Phoenix Refineries.	306FW	CAVU	0-4-30
	BOMBS	BOMBING	VICTORIES
	500lb GP	Visual	0-0-0
	FUSES		FLAK
	.1 Nose, .025 Tail		I-A-H
19 Ploesti, ROU - Marshalling Yard 6/51944	ESCORT	WEATHER	LOSSES
Photos show many direct hits on marshalling yard, factory area and coal storage yard East of the M/Y. Roundhouse took several direct hits and explosions were seen in the area adjacent to the roundhouse. Tracks have been cut.	14FG	LIGHT HAZE	
	BOMBS	BOMBING	VICTORIES
	FUSES		FLAK
20 Bucharest, ROU 7/5/1944	ESCORT	WEATHER	LOSSES
Photos show target area was well covered. Around 20 hits covered the round house near the aiming point. Forty hits in new siding, which was filled with rolling stock. 20 hits along South side of A/D. At least 10 tank wagons burning. Area surrounding target area well covered.	P-47s	9/10	0-0-3
	BOMBS	BOMBING	VICTORIES
	500lb GP	Visual	0-0-0
	FUSES		FLAK
	.1 Nose, .025 Tail		M-I-H

21 Weiner Neustadt, AUT 10/5/1944	ESCORT P-47s	WEATHER Haze	LOSSES 1-9-17
Photos show hits Southeast of dispersal area and two direct hits on 2 hangers South of the field. Hits scored across North dispersal field with undetermined number of A/C destroyed. Ammunition area received direct hits causing several explosions. Boiler house and locomotive works show fire and smoke. Scattered hits fell on flak battery.	BOMBS	BOMBING	VICTORIES
	120lb Frags	Visual	9-7-4
	FUSES		FLAK I-A-H

22 Marina di Carrera, ITA - Marshalling Yard 12/5/1944	ESCORT 52FG	WEATHER 9/10	LOSSES 0-0-0
29 planes dropped 72.5 tons of 500lb GP bombs. Photos show hits on Aluminum Potassium Plant. Railroad was cut near this point and just West of the marshalling yard.	BOMBS	BOMBING	VICTORIES
	500lb GP		0-0-0
	FUSES .1 Nose, .025 Tail		FLAK None

23 Imola, ITA - Marshalling Yard 13/5/1944	ESCORT 82FG	WEATHER 5/10	LOSSES 0-0-0
Bomb strike photos show at least 6 direct hits on marshalling yard among freight wagons. Direct hits on two large buildings, on 2 factory buildings, and on tracks East of the main yard. Tracks are believed cut in at least 12 places.	BOMBS	BOMBING	VICTORIES
	500lb GP		0-0-0
	FUSES .1 Nose, .025 Tail		FLAK

24 Padua, ITA 14/5/1944	ESCORT P-38s	WEATHER CAVU	LOSSES 0-4-0
71 A/C dropped 163.5 tons of 500lb bombs On this target. Bomb pattern continued through M/Y, scoring at least 30 direct hits on tracks, and destroying goods wagons. Good concentration also scored on choke points. A/D also received direct hits.	BOMBS	BOMBING	VICTORIES 0-0-0
	500lb GP		
	FUSES .1 Nose, .025 Tail		FLAK I-A-H

25 Elba Island, Corsica - Porto Ferrajo Steel Mill and Harbor 17/5/1944

Over 300 direct hits within 2000 feet of aiming point. Steel works building received at least 50 direct hits. Explosions were observed in the blast furnaces. Direct hits were scored on the jetty. A 220 foot M/V received a direct hit and a number of small vessels were sunk.

ESCORT	WEATHER	LOSSES
P-38s, P-51s	CAVU	0-0-0
BOMBS	BOMBING	VICTORIES
500lb GP		0-0-0
FUSES		FLAK
.1 Nose, .025 Tail		M-I-M

26 Belgrade, YUG - Zemun Airdrome 18/5/1944

Due to complete cloud cover at target no results were observed.

ESCORT	WEATHER	LOSSES
None	10/10	0-0-2
BOMBS	BOMBING	VICTORIES
500lb GP	Visual	0-0-0
FUSES		FLAK
.1 Nose, .025 Tail		S-I-H

27 Recco, ITA - Road Viaduct 19/5/1944

Large bomb crater and two near misses severed the track 350 feet from the tunnel entrance. Two direct hits on the West entrance of the arched bridge.

ESCORT	WEATHER	LOSSES
P-38s, P-51s	9/10	0-0-0
BOMBS	BOMBING	VICTORIES
1000lb GP	Visual	0-0-0
FUSES		FLAK
.1 Nose, .025 Tail		None

28 Piombino, ITA - Harbor Area 22/5/1944

Target completely covered, blanketing coke ovens, blast furnace, RR yards, fuel storage tanks, power stations, tin plate mills, and steel mills. Entire industrial area completely destroyed.

ESCORT	WEATHER	LOSSES
P-51s	CAVU	0-1-15
BOMBS	BOMBING	VICTORIES
1000lb GP		0-0-0
FUSES		FLAK
Prin		M-A-H

29 Subiaco, ITA - Road Junction 23/5/1944

Good concentration on aiming point and another concentration on road fork 2 1/2 miles from aiming point.

ESCORT	WEATHER	LOSSES
52FG	CAVU	0-0-0
BOMBS	BOMBING	VICTORIES
500lb GP		0-0-0
FUSES		FLAK
.1 Nose, .025 Tail		None

30 Weiner Neustadt, AUT - Wollersdorf Airdrome 24/5/1944	ESCORT P-38s	WEATHER LIGHT HAZE	LOSSES 2-3-20
Photo through a hole in the clouds show about 80 hits through the center of the landing area destroying 1 A/C and probably 3 more.	BOMBS 500lb GP	BOMBING	VICTORIES 1-3-3
	FUSES .1 Nose, .025 Tail		FLAK

31 Carnoules, FRA - Marshalling Yard 25/5/1944	ESCORT P-51s	WEATHER CAVU	LOSSES 0-0-0
Hits on Round House, circular storage yard and main M/Y. String of bombs across West choke point, cutting it in two places. Hits on main tracks cutting them in three places.	BOMBS 500lb GP	BOMBING	VICTORIES 0-0-0
	FUSES .1 Nose, .025 Tail		FLAK None

32 Lyon Vaise, FRA - Marshalling Yard 26/5/1944	ESCORT P-38s, P-51s	WEATHER CAVU	LOSSES 0-0-0
Completely covered target area. At least 50% of wagons in M/Y were destroyed or damaged. All RR shop buildings received at least 5 direct hits. All car shops and factories in target area destroyed or damaged.	BOMBS 500lb GP	BOMBING	VICTORIES 0-0-0
	FUSES .1 Nose, .025 Tail		FLAK None

33 Salon De Provence, FRA - Airdrome 27/5/1944	ESCORT P-38s, P-51s	WEATHER LIGHT HAZE	LOSSES 1-1-25
2 direct hits and 3 near misses in NE hanger, 4 direct hits and 6 near misses on larger NW hanger. 2 small center hangers were blanketed by bombs. Administrative building containing two control towers sustained 5 direct hits. Barracks building in NW corner of field also hit. Landing area, hanger foundations, dispersal area also hit.	BOMBS 500lb GP	BOMBING	VICTORIES 0-0-0
	FUSES .1 Nose, .025 Tail		FLAK M-A-H

34 Weiner Neustadt, AUT - Werke #1 Nord Airdrome 29/5/1944

Photos show hangers and factory area completely saturated with bursts. 7 S/E A/C in front of hangers, and 10 other A/C in various parts of the airdrome were destroyed. At least 7 additional aircraft were damaged. Hangers received direct hits. Factory area received direct hits. RR junction and additional hits cut EBENFURTH line. Bombs covered barracks SE of Wiener Neustadt A/D.

ESCORT	WEATHER	LOSSES
P-38s	CAVU	0-0-7
BOMBS	BOMBING	VICTORIES
500lb GP		3-0-0
FUSES		FLAK
.1 Nose, .025 Tail		M-A-H

35 Wels, AUT - Aircraft Factory 30/5/1944

Concentrated pattern of hits covering the entire target. Bombs from 484th attack created so much smoke that detailed interpretation of photos from bombs of the 461st and 451st were impossible. The later bombs were seen going into the smoke covered target.

ESCORT	WEATHER	LOSSES
P-38s, P-51s	CAVU	0-1-4
BOMBS	BOMBING	VICTORIES
100lb Incind		0-0-0
FUSES		FLAK
Prin		None

36 Ploesti, ROU - Concordia Vega Oil Refinery 31/5/1944

Two direct hits on NE Distillation Plant & 3 near misses. Boiler House received 2 direct hits and a cracking plant in the west central area received 2 direct hits and a near miss. RR Station and yards South of Refinery have received strings of bombs and storage tanks in SW corner of Refinery have been hit, Scattered hits in built up area South of target. Numerous smoke pots are in effective use.

ESCORT	WEATHER	LOSSES
P-38s	CAVU	2-5-26
BOMBS	BOMBING	VICTORIES
500lb GP		0-0-0
FUSES		FLAK
.1 Nose, .025 Tail		I-A-H

Four 461st ships were lost to enemy flak, one to enemy fighters, and two others to engine trouble. Twenty nine men were killed in action and another 24 were wounded. Thirty five became prisoners of war.

June 1944

There were 468 sorties flown with 65 early returns. The group dropped 1,369 tons of bombs, almost all of which were on briefed targets, including Germany for the first time. Targets were primarily oil refineries and oil storage facilities plus some marshalling yards. These targets directly reflect the stated strategy of the Fifteenth Air Force in attacking oil production and transportation networks. Ten enemy aircraft were destroyed with seven probables and another three damaged.

37 Szolnok, HUN - Marshalling Yard 2/6/1944	ESCORT 82FG	WEATHER CAVU	LOSSES 0-0-0
Bomb strike photos show six strings across M/Y containing over 1500 mixed goods wagons. Eight explosions are visible among the rolling stock. Two locomotive sheds received hits. The Eastern choke point was smothered by a large concentration and rolling stock and trackage suffered considerable damage.	BOMBS 500lb GP	BOMBING Visual	VICTORIES 0-0-0
	FUSES .1 Nose, .025 Tail		FLAK S-I-M

38 Orelle, ITA - Railroad Bridges/Viaducts 4/6/1944	ESCORT 52FG	WEATHER CAVU	LOSSES 0-0-0
Photos show the RR tracks were cut 5,000 feet East of ST MICHELE - some other bombs hit the town of ST MICHELE and several scored hits on the road West of town. Another string hit the tracks at the East edge of the town with a possible hit on a South approach to a highway bridge over the river.	BOMBS 1000lb GP	BOMBING Visual	VICTORIES 0-0-0
	FUSES .1 Nose, .025 Tail		FLAK None

39 Fornovo di Taro, ITA - Railroad Bridge 5/6/1944	ESCORT 14FG	WEATHER CAVU	LOSSES 0-0-0
Bomb strike photos show a concentration cutting the tracks on the south side of town with hits 400 feet North of the rail and road bridges. There were several hits on the east bank of the river and scattered bursts in the open land just South of the RR marshalling yard.	BOMBS 1000lb GP	BOMBING Visual	VICTORIES 0-0-0
	FUSES .1 Nose, .025 Tail		FLAK None

40 Ploesti, ROU - Xenia Oil Refinery 6/6/1944	ESCORT 1, 14, 82 FGs	WEATHER CAVU	LOSSES 0-1-18
Photos show target to be heavily obscured by smoke pots and cloud making bomb strike assessment impossible.	BOMBS 500 lb. GP	BOMBING Visual	VICTORIES 1-0-0
	FUSES .1 Nose, .025 Tail		FLAK I-A-L

41 Antheor, FRA - Railroad Viaduct 7/6/1944	ESCORT	WEATHER	LOSSES
Strikes form a good concentration in the target area. 10 hits bracketed the viaduct from the center to the north end. A string of hits cut the track 1200 feet south. 5 hits or near misses at the north approach cut tracks and possibly damaged a string of wagons along the north portion of a span.	P-38sNone	CAVU	0-3-15
	BOMBS	BOMBING	VICTORIES
	1000lb Armor	Visual	0-0-0
	FUSES		FLAK
	.1 Nose, .025 Tail		M-A-H

42 Munich, DEU - West Marshalling Yard 9/6/1944	ESCORT	WEATHER	LOSSES
Bomb strike photos (obtained through a break in the clouds) show a cluster of hits smothering the RR junction and choke point west of main M/Y and a number of hits on the double tracks. There were over 20 hits on siding and tracks and a number of hits on railroad buildings at the NE point of the RR yard.	P-38s	10/10	0-0-14
	BOMBS	BOMBING	VICTORIES
	500lb GP	Pathfinder	0-0-1
	FUSES		FLAK
	.1 Nose, .025 Tail		M-I-H

43 Porto Marghera, ITA - Oil Refinery 10/6/1944	ESCORT	WEATHER	LOSSES
Large concentration of hits in the assigned target area. Many large fires are visible and four large storage tanks are seen to explode. 22 direct hits are seen on the Sevetra Coke and By-Product plant west of the target. Small R.R. sidings at the plant received at least 6 direct hits damaging or destroying at least 50 wagons.	14FG	CAVU	0-0-11
	BOMBS	BOMBING	VICTORIES
	500lb GP	Visual	0-0-0
	FUSES		FLAK
	.1 Nose, .025 Tail		M-I-H

44 Giurgiu, ROU - Oil Storage 11/6/1944	ESCORT	WEATHER	LOSSES
Strike photos from 49th BW show very highly concentrated pattern of hits in assigned target area. Seven direct hits fell on the oil loading quay. 2 direct hits were made on oil storage tanks and one large explosion was seen. The oil pumping station and 20 oil tanks were probably damaged by numerous near misses. Two direct hits were scored on the RR junction.	1FG, 14FG	CAVU	2-0-11
	BOMBS	BOMBING	VICTORIES
	250lb GP	Visual	5-4-1
	FUSES		FLAK
	.1 Nose, .025 Tail		M-A-H

45 Porto Marghera, ITA - Oil Storage 13/6/1944	ESCORT	WEATHER	LOSSES
Bomb strike photos are badly hazed rendering accurate assessment impossible. 3 hits are visible in the oil storage depot. The majority of bombs fell on the Sava Aluminum Works and there were 5 hits on the Savetero Works.	None	3/10	0-0-9
	BOMBS	BOMBING	VICTORIES
	500lb GP	Pathfinder	0-0-1
	FUSES		FLAK
	.1 Nose, .025 Tail		M-I-H

46 Szony, HUN - Oil Storage 14/6/1944

Entire target area was well covered with main concentration in the boiler and pumping house area. 3 direct hits were scored on the Pumping House and many direct hits and near misses were scored on the Boiler House, Distillation House, running tanks and in the oil storage tank area. The rail sidings that serviced the refinery also received hit cutting the tracks.

ESCORT	WEATHER	LOSSES
325 FG	CAVU	0-0-0
BOMBS	BOMBING	VICTORIES
250lb GP	PFF	0-0-0
FUSES		FLAK
.1 Nose, .025 Tail		S-I-H

47 Trieste, ITA - Oil Storage 22/6/1944

43 B-24s dispatched to primary target found it to be completely overcast and unable to locate objective jettisoned or returned bombs to base. No enemy aircraft - No flak - No claims.

ESCORT	WEATHER	LOSSES
None	10/10	0-0-0
BOMBS	BOMBING	VICTORIES
250lb GP	Did Not Bomb	0-0-0
FUSES		FLAK
.1 Nose, .025 Tail		S-I-H

48 Giurgiu, ROU - Oil Storage 23/6/1944

461st Bomb Group photos show a well concentrated pattern. Storage tanks blanketed North of RR siding, with 50 hits within a 1000 foot radius. One large tank in center of yard was hit, causing a large explosion. An adjoining tank was probably also hit causing a second explosion.

ESCORT	WEATHER	LOSSES
P-47, P-51	3/10	0-0-29
BOMBS	BOMBING	VICTORIES
1000lb RDX	PFF	0-0-0
FUSES		FLAK
.1 Nose, .025 Tail		I-A-H

49 Avignon, FRA - East Marshalling Yard 25/6/1944

String of hits on goods wagons at south end of yard. 15 hits among goods wagons and tracks in center area of the yard. Double track line to CAVAILLON hit in four places. Roads to north and south parts of yard received hits and near misses were scored on turn-table.

ESCORT	WEATHER	LOSSES
31FG	CAVU	0-0-0
BOMBS	BOMBING	VICTORIES
500lb GP	Visual	0-0-0
FUSES		FLAK
.1 Nose, .025 Tail		None

50 Korneuburg, AUT - Oil Refinery 26/6/1944

Hits blanketed a large barracks area Northeast of the target scoring 15 D/H's and near misses on the buildings. Several strikes were scored on trackage and on two highways to the East of the target.

ESCORT	WEATHER	LOSSES
82FG	CAVU	1-0-7
BOMBS	BOMBING	VICTORIES
500lb GP	Pathfinder	3-1-0
FUSES		FLAK
.1 Nose, .025 Tail		M-I-H

41 Bucharest, ROU - Chitila Marshalling Yard 28/6/1944	ESCORT	WEATHER	LOSSES
Strike photos show a concentration of hits in the dispersal area at the North edge of GIULESTI A/D. Several hits at West edge of barracks and car repair buildings. A second concentration and 5 hits on the large car repair shop. Some hits in the residential area.	52FG	CAVU W/ HAZE	0-0-3
	BOMBS	BOMBING	VICTORIES
	500lb GP	Pathfinder	1-2-0
	FUSES		FLAK
	.1 Nose, .025 Tail		I-I-H

52 Blechhammer, DEU - Synthetic Oil Plant 30/6/1944	ESCORT	WEATHER	LOSSES
38 A/C took off to bomb BLECHHAMMER. There were 5 early returns. However, a complete undercast forced the Group to turn back. Of the 33 remaining aircraft 31 returned their bombs to Torretta and 2 dropped their bombs in the Adriatic.	P-38	10/10	0-0-2
	BOMBS	BOMBING	VICTORIES
	500lb GP	Return bombs	0-0-0
	FUSES		FLAK
	.1 Nose, .025 Tail		M-A-M

The group lost two planes to fighters and one to flak. Three men were killed and twelve wounded. Seventeen became POWs. Six replacement crews joined the 461st.

July 1944

The group flew 21 combat missions with 630 sorties and 68 early returns. Gunners shot down 43 enemy aircraft, probably destroyed another 23, and damaged 9 more.

53 Budapest, HUN - Rakos Marshalling Yard 2/7/1944	ESCORT	WEATHER	LOSSES
Over 100 hits concentrated on the small East M/Y, car shops, sheds, and tracks. 2 direct hits on the East roundhouse and 1 direct hit on the West roundhouse with 5 near misses. 15 direct hits on S.E. area of railway repair shops North of target. Highway overpass 700 feet from target received direct hit. All lines blocked and extensive damage to tracks and rolling stock.	31, 51, 325 FG	CAVU	0-8-0
	BOMBS	BOMBING	VICTORIES
	500lb GP	Visual	0-0-0
	FUSES		FLAK
	.1 Nose, .025 Tail		H-I-A

54 Bucharest, ROU - Mogasaia Oil Storage 3/7/1944

Iron Gates Rail Embankment - 4 direct hits are seen cutting through tracks at the East point of the M/Y, 3 direct hits on the Southeast part of the M/Y and several North of the yards. The adjacent highway was cratered by several hits.

ESCORT	WEATHER	LOSSES
52FG, 82FG	4/10	1-4-0
BOMBS	BOMBING	VICTORIES
500lb GP	Visual	0-0-0
FUSES		FLAK
.1 Nose, .025 Tail		H-M-I

55 Beziers, FRA - Marshalling Yard 5/7/1944

Many hits in East rail yard which contained many goods wagons. Many direct hits on track between locomotive repair shop and car repair shops. Large pattern across yard West of locomotive shop. Hits and near misses on turn-table. Direct hits on factory building North of locomotive shop causing explosions and fire. Oil storage tanks at South end of yard hit and burning.

ESCORT	WEATHER	LOSSES
31FG	3/10	0-0-0
BOMBS	BOMBING	VICTORIES
500lb GP	Visual	0-0-0
FUSES		FLAK
.1 Nose, .025 Tail		S-I-H

56 Aviano, ITA, Oil Storage 6/7/1944

Target completely smothered by a concentration of hits. First bombs fell SE with bulk of pattern on and around buildings of the target area. 3 barrack buildings to SE received direct hits. Hits on the pipeline and storage buildings causing two large explosions throwing up much flame and smoke. Over 80% of hits within 1000' radius of aiming point.

ESCORT	WEATHER	LOSSES
82FG	CAVU	0-0-0
BOMBS	BOMBING	VICTORIES
500lb GP	Visual	0-0-0
FUSES		FLAK
.1 Nose, .025 Tail		None

57 Blechhammer, DEU - Synthetic Oil Plant 7/7/1944

Bomb Strike photo shows first hits fell West of target. Bomb pattern cuts across M/Y at KANDRZAN showing 15 direct hits cutting tracks just South of choke point and continuing on to target. Numerous smoke pots are visible emitting white smoke over target area and penetrated by billows of black smoke, indicated probable oil fires.

ESCORT	WEATHER	LOSSES
306BW	4/10	0-0-21
BOMBS	BOMBING	VICTORIES
500lb GP	Pathfinder	5-4-1
FUSES		FLAK
.1 Nose, .025 Tail		I-A-H

58 Korneuburg, AUT - Oil Refinery 8/7/1944	ESCORT 1FG	WEATHER CAVU	LOSSES 0-0-17
Many strikes smothered the Refinery and caused two large explosions and numerous fires. Approximately 15 hits were scored on the quay side damaging oil loading facilities and tracks. 4 direct hits fell on storage siding cutting tracks and damaging tank wagons. The through line to VIENNA was cut by a number of direct hits.	BOMBS 500lb GP	BOMBING Visual	VICTORIES 0-0-0
	FUSES .1 Nose, .025 Tail		FLAK I-A-H

59 Toulon, FRA - Submarines in Harbor 11/7/1944	ESCORT 332FG	WEATHER CAVU	LOSSES 0-0-10
Concentration of hits on telegraph cable factory at West shore of LASEYNE BAY. 11 hits in M/Y NW of Bay. 15 strikes on and around LAGOUBRAN jetties and oil stores and in barracks and repair shops area North of jetties. 8 hits and 4 near misses on dry docks containing 3 large and 2 small submarines. 18 hits on M/Y and repair shops North of target. Train stations also hit.	BOMBS 500lb GP	BOMBING 500lb GP	VICTORIES 0-0-0
	FUSES .1 Nose, .025 Tail		FLAK I-A-H

60 Nimes, FRA - Marshalling Yard 12/7/1944	ESCORT 325FG, 332FG	WEATHER CAVU	LOSSES 4-0-10
Tracks to TARASCON cut in 6 places. The heaviest concentration fell on goods wagons and tracks bracketing the East area of the yard. Rail sidings to North edge of yard received 5 direct hits. Several hits cut rail line to AVIGNON. Main center yard received 18 direct hits. 6 hits on car repair shops, and 10 hits in shop area along South edge of central yard.	BOMBS 500lb GP	BOMBING Visual	VICTORIES 7-0-0
	FUSES .1 Nose, .025 Tail		FLAK S-I-H

61 Petfurdo, HUN - Oil Refinery 14/7/1944

Concentration on the N part of the Refinery with direct hits on the power plant, gas generator, the stabilization and cracking plant buildings. The compression plant in the center of the target suffered at least 2 direct hits and several near misses. Several direct hits were scored on the underground storage area in the NE area of the refinery. Direct hits on goods wagons.

ESCORT	WEATHER	LOSSES
82FG	CAVU	0-0-0
BOMBS	**BOMBING**	**VICTORIES**
1000lb GP	Visual	0-0-0
FUSES		**FLAK**
.1 Nose, .025 Tail		S-I-H

62 Ploesti, ROU - Creditul Minier Oil Refinery 15/7/1944

Target area blanketed by hits. Over 125 hits are visible on the refinery, including 1 complete string on the South half. One large explosion was shown in the tank car storage yard in the South part of the target. 18 hits in the partly filled tank car storage area. 21 hits on the choke point of the M/Y East of the target cutting the through lines to BUCHAREST.

ESCORT	WEATHER	LOSSES
82FG	6/10	1-0-13
BOMBS	**BOMBING**	**VICTORIES**
500lb GP	Pathfinder	0-0-0
FUSES		**FLAK**
.1 Nose, .025 Tail		I-A-H

63 Weiner Neustadt, AUT - Aircraft Engine Factory 16/7/1944

A heavy concentration is seen in the Southwestern part of the factory area. 5 direct hits were scored on two large machine and assembly buildings and two near misses on a third building. 2 small shop buildings at the West edge of the target area also received direct hits. A large cluster fell on the engine area with 10 direct hits. Several hits were seen on rail sidings to the Northeast.

ESCORT	WEATHER	LOSSES
1FG, 14FG	9/10	0-0-16
BOMBS	**BOMBING**	**VICTORIES**
1000lb RDX	Pathfinder	0-0-0
FUSES		**FLAK**
.1 Nose, .025 Tail		I-A-H

64 Friedrichshafen, DEU - Manzell Dornier Werke 18/7/1944

Main concentration of strikes on the North edge of the target. Direct hits were scored on the Electrolysis Shop, the multi-function machine shop, 2 large warehouses in the East target area and trackage at the North edge of the Works.

ESCORT	WEATHER	LOSSES
1 FG	CAVU	0-2-10
BOMBS	**BOMBING**	**VICTORIES**
1000lb RDX	Pathfinder	0-0-0
FUSES		**FLAK**
.1 Nose, .025 Tail		I-A-H

65 Schleissheim Airdrome Installations 19/7/1944

Compact pattern on target, with direct hits on many hangers and on East end of long building adjoining center hanger. E.-W. runway pitted with direct hits. Large shops on West half of target and small shop buildings were directly hit. 2 A/C destroyed, 2 damaged and several visible inside damaged hangers were probably destroyed. Main assembly building damaged by near misses.

ESCORT	WEATHER	LOSSES
332 FG	5/10	0-0-18
BOMBS	BOMBING	VICTORIES
1000lb GP	Pathfinder	2-0-2 (Grd)
FUSES		FLAK
.1 Nose, .025 Tail		I-A-H

66 Brux, CZE - Synthetic Oil Refinery 21/7/1944

Bombs fell to the East of the target area with no military damage evident.

ESCORT	WEATHER	LOSSES
14FG, 82FG	5/10	0-2-0
BOMBS	BOMBING	VICTORIES
100lb GP	Pathfinder	0-0-0
FUSES		FLAK
.1 Nose, .025 Tail		I-A-H

67 Ploesti, ROU - Romano Americana Oil Refinery 22/7/1944

Bomb strike photos show target generally obscured by cloud, smoke screen and heavy black smoke emanating from several fires in the target area. First Phase Photo Reconnaissance show hits were scored on the West end of the Romano Americana Refinery where fires were started among the large storage tanks at least three of which are burning.

ESCORT	WEATHER	LOSSES
P-38's	6/10	4-0-19
BOMBS	BOMBING	VICTORIES
1000lb GP	Pathfinder	0-0-0
FUSES		FLAK
.1 Nose, .025 Tail		I-A-H

68 Pljevlja, Prejepolje, Sjenica, Andrijevica, YUG 24/7/1944

Pljevlja - 30 plus hits directly on target. Buildings heavily hit and highway cut. Prejepolje - 6 direct hits on buildings and main highway in center of target area. Sjenica - Crews report excellent concentration of bombs on target. Clouds obscured photos. Andrijevica - Crews report a few scattered hits in the SW area of the target. Clouds obscured photos.

ESCORT	WEATHER	LOSSES
P-51s	3/10	0-0-0
BOMBS	BOMBING	VICTORIES
1000lb GP	Visual	0-0-0
FUSES		FLAK
.1 Nose, .025 Tail		None

Mission	Escort / Bombs	Weather / Bombing	Losses / Victories / Flak
69 Linz, AUT - Herman Goering Tank Work 25/7/1944	ESCORT 306BW	WEATHER CAVU	LOSSES 15-3-4
Effectiveness of bomb strikes could not be determined as no photo planes returned from the mission.	BOMBS 1000lb GP	BOMBING Pathfinder	VICTORIES 31-19-7
	FUSES .1 Nose, .025 Tail		FLAK I-A-H
70 Pec, YUG - City Proper 27/7/1944	ESCORT P-51s	WEATHER CAVU	LOSSES 0-0-0
24 B-24s were dispatched. A/C dropped 74 tons (148 X 1000lb GP) from 18,000 feet. Bombing was visual. Bombs dropped South of target area.	BOMBS 1000lb GP	BOMBING Pathfinder	VICTORIES 0-0-0
	FUSES .1 Nose, .025 Tail		FLAK None
71 Phlorina, GRC - RR Station/Marshalling Yard 28/7/1944	ESCORT 325th	WEATHER CAVU	LOSSES 2(NC)-0-0
Bomb strike photos show a pattern of close strikes crossing the railhead and the trackage leading into it. 15 hits are seen on track and rolling stock, with over 40 hits in the immediate area. The East end of town and hutment area to the NE received direct hits.	BOMBS 1000lb GP	BOMBING Visual	VICTORIES 0-0-0
	FUSES .1 Nose, .025 Tail		FLAK None
72 Budapest, HUN - Duna Aircraft Factory 30/7/1944	ESCORT P-38s	WEATHER 2/10	LOSSES 0-0-12
Concentrated pattern on East side of target edge of Factory building. Some hits in the N.E. barracks area and 5 direct hits cutting rail lines. 20 direct hits and several near misses on assembly buildings and test beds.	BOMBS 500lb RDX	BOMBING Visual	VICTORIES 0-0-0
	FUSES .1 Nose, .025 Tail		FLAK M-A-H
73 Bucharest, ROU - Prahova Oil Refinery 31/7/1944	ESCORT 52FG	WEATHER 5/10	LOSSES 0-0-5
Several hits in the building area West of the PRAHOVA Refinery; 5 direct hits on railroad sidings at South edge; 7 on railroad junction and 10 on railroad North of the refinery. Railroad to FUNDULEA was cut by 7 direct hits. 15 hits were scored on GRIVITA M/Y and rolling stock: 1 on locomotive depot. Balance of strikes were cloud obscured.	BOMBS 500lb RDX	BOMBING Visual	VICTORIES 0-0-0
	FUSES .1 Nose, .025 Tail		FLAK M-I-H

The 461st lost 24 ships with another 19 damaged. The group suffered 57 killed and 43 wounded. Another 104 became prisoners of war.

Twenty six new crews joined the group during July.

August 1944

The group flew 20 combat missions during August. Of the 586 sorties, 54 were early returns. Of the 1,220 tons of bombs dropped, 1,059 were dropped on targets. Bombing accuracy for the month was erratic going as low as 0% and as high as 73.9 for an average of 42.2 per cent. As a result the group ranked thirteenth in the Air Force for the month. Gunners claimed sixteen fighters knocked down with six more probables, and one damaged.

74 Avignon, FRA - Railroad Bridge 2/8/1944	ESCORT	WEATHER	LOSSES
South bridge received direct hits at South approach along river bank and others near center. Many near misses bracketed the bridge.	31FG, 325FG	1/10	0-0-2
	BOMBS	BOMBING	VICTORIES
	500lb GP	Visual	0-0-0
	FUSES		FLAK
	.1 Nose, .025 Tail		None

75 Friedrichshafen, DEU - Zahnradfabrik A/C Factory 3/8/1944	ESCORT	WEATHER	LOSSES
Bomb Strike Photos are too heavily cloud covered to permit interpretation.	52FG, 332FG	8/10	1-0-4
	BOMBS	BOMBING	VICTORIES
	500lb RDX	Pathfinder	0-0-0
	FUSES		FLAK
	.1 Nose, .025 Tail		S-I-H

76 Miramas, FRA - Marshalling Yard 6/8/1944	ESCORT	WEATHER	LOSSES
Direct hit on West choke point and line leading to ARLES. Rolling stock and tracks in central yard and sidings were covered by many hits. Sulphuric acid plant to South of yard hit and sidings at North end received many strikes among goods wagons. Tracks hit at triangular junction to AVIGNON.	None	CAVU	0-0-6
	BOMBS	BOMBING	VICTORIES
	500lb GP	Visual	0-0-0
	FUSES		FLAK
	.1 Nose, .025 Tail		M-I-H

77 Blechhammer, DEU - South Oil Refinery 7/8/1944			
Smoke from oil fires in northwest corner of area, and concentration of hits in northwest corner. Bomb pattern covered RR sidings to the East of Gas Plants and Boiler House extending into Gas Generating Plant. Bombs also carried into labor camp North of the target. Direct hits on buildings of Gas Plants, Boiler House, and adjacent buildings.	ESCORT 52FG, 1FG	WEATHER CAVU	LOSSES 1-0-12
	BOMBS 500lb RDX	BOMBING Pathfinder	VICTORIES 0-0-0
	FUSES .1 Nose, .025 Tail		FLAK None

78 Almasfuzito, HUN - Oil Refinery 9/8/1944			
Compact pattern began on South perimeter of refinery and continued to center of target. The South RR sidings were well covered with many strikes damaging many tanks and goods wagons. The Southern Lubrication Oil Plant received 5 hits and the final products storage tanks received three hits causing a huge explosion. 75% of target in smoke and flames.	ESCORT 31FG, 52FG	WEATHER CAVU	LOSSES 0-0-6
	BOMBS 500lb GP	BOMBING Visual	VICTORIES 0-0-0
	FUSES .1 Nose, .025 Tail		FLAK M-I-H

79 Ploesti, ROU - Xenia Oil Refinery 10/8/1944			
Bomb strike photos show several explosions in the target area. Further assessment is impossible due to the heavy smoke screen.	ESCORT 52FG	WEATHER CAVU	LOSSES 0-1-12
	BOMBS 500lb RDX	BOMBING Pathfinder	VICTORIES 0-0-0
	FUSES .1 Nose, .025 Tail		FLAK I-A-M

80 Genoa, ITA - Gun Positions 12/8/1944			
The Group dropped 57.5 tons of bombs scoring only near misses. One aircraft lost to flak, another aircraft is missing.	ESCORT 82FG	WEATHER CAVU	LOSSES 1-0-9
	BOMBS 500lb RDX	BOMBING Visual	VICTORIES 0-0-0
	FUSES .1 Nose, .025 Tail		FLAK M-A-H

81 Genoa, ITA - Gun Positions 13/8/1944

The 461st was assigned to drop 54 tons of bombs on gun positions in Genoa, Italy, but failed to hit the target.

ESCORT	WEATHER	LOSSES
None	3/10	0-0-0
BOMBS	**BOMBING**	**VICTORIES**
500lb RDX	Visual	0-0-0
FUSES		**FLAK**
.1 Nose, .025 Tail		M-A-H

82 Frejus, FRA - Gun Positions 14/8/1944

Group was assigned to drop 40.5 tons of bombs on one position and 22 tons on another gun position. While the first target was missed, the Group scored two direct hits on the second target causing a large explosion.

ESCORT	WEATHER	LOSSES
None	CAVU	0-0-0
BOMBS	**BOMBING**	**VICTORIES**
1000lb GP	Visual	0-0-0
FUSES		**FLAK**
.1 Nose, .025 Tail		None

83 Frejus, FRA - Beach Defenses 15/8/1944

Good to excellent coverage. Damage to nearby roads and railroads, and to four of six target areas.

ESCORT	WEATHER	LOSSES
None	CAVU W/ HAZE	0-0-0
BOMBS	**BOMBING**	**VICTORIES**
100lb GP	Visual	0-0-0
FUSES		**FLAK**
No Delay Fuses		None

84 Ploesti, ROU - Romano Americana Oil Refinery 17/8/1944

Bomb Strike photos from 49th Wing show direct hits on Tank farm, Oil tanks, Distillation and Cracking Plants, Workshops, Storage and Administrative building with one string of bombs across center of refinery and four large fires started. 1 A/C lost to flak.

ESCORT	WEATHER	LOSSES
31FG	CAVU	1-0-15
BOMBS	**BOMBING**	**VICTORIES**
500lb RDX	Visual	0-0-0
FUSES		**FLAK**
.1 Nose, .025 Tail		I A H

85 Alibunar, YUG - Airdrome 18/8/1944

Strings fell on Northeast side of field through dispersal area, and from center of landing ground to Northwest. Dispersal area at South end also covered. 30 E/A damaged and 10 E/A probably damaged.

ESCORT	WEATHER	LOSSES
82FG	3/10	0-0-0
BOMBS	**BOMBING**	**VICTORIES**
Frag Clusters	Visual	0-0-0
FUSES		**FLAK**
Instantaneous		None

86 Szolnok, HUN - Airdrome20/8/1944

Strings of bombs falling from North of hangers South across the A/D. South dispersal area was covered with an even pattern. Direct hits severely damaged or destroyed 16 A/C and near misses probably damaged 10. Approximately 95 aircraft were seen on the ground.

ESCORT	WEATHER	LOSSES
None	HAZE	0-0-0
BOMBS	BOMBING	VICTORIES
Frag Clusters	Visual	9-0-11
FUSES		FLAK
Instantaneous		S-I-M

87 Vienna, AUT - Lobau Oil Storage 22/8/1944

Close pattern of hits on center of plant. Many hits on underground storage tank and several hits on partially filled oil tank storage area. Smoke prevented further assessment.

ESCORT	WEATHER	LOSSES
1FGm 14FG	CAVU	1-3-12
BOMBS	BOMBING	VICTORIES
1000lb GP	Visual	6-1-0
FUSES		FLAK
.1 Nose, .025 Tail		I-A-H

88 St. Polten, AUT - Markersdorf Airdrome 23/8/1944

Bomb pattern started 1,000 feet SW of the target and strung NE across the airdrome to hanger installations with even coverage of the entire target. 15 E/A were destroyed or damaged on the ground. Approximately 30 E/A were seen. Some bombs fell approximately 9 miles WNW of the airdrome.

ESCORT	WEATHER	LOSSES
52, 325, 322 FGs	CAVU	1-2-12
BOMBS	BOMBING	VICTORIES
Frag Clusters	Pathfinder	4-2-3
FUSES		FLAK
Instantaneous		None

89 Ferrara, ITA - Railroad Bridge 24/8/1944

9 direct hits and three near misses on elevated railroad bridge approach, probably damaging first span. Industrial area South of river received approximately 60 hits starting two large fires.

ESCORT	WEATHER	LOSSES
None	CAVU W/ HAZE	1-1-22
BOMBS	BOMBING	VICTORIES
1000lb GP	Visual	0-0-0
FUSES		FLAK
.1 Nose, .025 Tail		I-A-H

Mission	Description	ESCORT	WEATHER	LOSSES	BOMBS	BOMBING	VICTORIES	FUSES	FLAK
90 Bucharest, ROU - Otopeni Airdrome 26/8/1944	Heavy concentration fell across South end of A/D with many hits on South end of North South runway and South perimeter track. Radio station, administration building and barracks were also hit. North South highway to PLOESTI received several hits.	325FG	CAVU	1-0-1	500lb GP	Visual	0-0-0	.1 Nose, .025 Tail	None
91 Venzone, ITA - Viaduct 27/8/1944	Dropped close pattern into bomb smoke from 451st bombs with four visible hits on RR viaduct.	2, 82 FGs	3/10	0-0-0	500lb GP	Visual	0-0-0	.1 Nose, .025 Tail	None
92 Szolnok, HUN - Szajol Railroad Bridge 28/8/1944	Bomb strike photos show the following: Very close pattern smothered the target area, with 4 possible direct hits and 11 near misses on the bridge, and 5 direct hits and 9 near misses on the approaches, cutting the tracks on both sides.	None	CAVU	0-0-0	1000lb GP	Visual	0-0-0	.1 Nose, .025 Tail	S-I-H
93 Szegog, HUN - Marshalling Yard 29/8/1944	Concentrated pattern across center of M/Y and warehouse area. Many direct hits on rolling stock and trackage starting fires. Three small warehouse buildings hit and one explosion seen. Through rails also hit.	None	CAVU	0-0-0	500lb GP	Visual	0-0-0	.1 Nose, .025 Tail	None

Eight aircraft were lost during the month with 12 men killed in action and 21 wounded. Fifty-three others became prisoners of war. Forty-seven new crews arrived at Torretta

September 1944

Only seven combat missions were flown during September because the group was assigned supply missions from 10 September through 22 September 1944. With an average bombing score of 57.9, the group finished the month ranked fourth in the Air Force. This was helped in no small part by a score of 92.2 on September 3, 1944 at Smederovo, Yugoslavia. No enemy aircraft were seen on any of the September missions.

<table>
<tr><td>94 Ferrara, ITA - Railroad Bridge
1/9/1944</td><td>ESCORT
14FG</td><td>WEATHER
CAVU W/ HAZE</td><td>LOSSES
0-0-4</td></tr>
<tr><td rowspan="2">Ferrara railroad bridge was not attacked due to 10/10 cloud cover. Instead, 26 461st BG aircraft attacked the Boari Pisani road bridge. Bomb strike photos (461st) show the bridges at BOARI PISANI across ADIGE River - majority of scattered pattern fell in fields North of the river. 1 direct hit probably cut highway bridge at North end. Several near misses on RR and highway.</td><td>BOMBS
500lb RDX</td><td>BOMBING
Visual</td><td>VICTORIES
0-0-0</td></tr>
<tr><td colspan="2">FUSES
.1 Nose, .025 Tail</td><td>FLAK
S-I-M</td></tr>
<tr><td>95 Mitrovica, YUG - Railroad Bridge
2/9/1944</td><td>ESCORT
None</td><td>WEATHER
CAVU</td><td>LOSSES
0-1-0</td></tr>
<tr><td rowspan="2">Closely concentrated pattern fell on West approach with 6 direct hits or near misses on rail line. 4 near misses are visible on track East of bridge. 2 direct hits on South approach of highway bridge South of target.</td><td>BOMBS
1000lb RDX</td><td>BOMBING
Visual</td><td>VICTORIES
0-0-0</td></tr>
<tr><td colspan="2">FUSES
.1 Nose, .025 Tail</td><td>FLAK
None</td></tr>
<tr><td>96 Smederovo, YUG - Ferry Slip
3/9/1944</td><td>ESCORT
None</td><td>WEATHER
CAVU</td><td>LOSSES
0-0-0</td></tr>
<tr><td rowspan="2">Close concentrated pattern of bombs fell on South Ferry Slip and RR lead-in. 2 direct hits and several near misses fell on ferry slip. 10 direct hits and several near misses fell on RR lead-in cutting it in several places. Barge at end of slip received three near misses.</td><td>BOMBS
500lb RDX</td><td>BOMBING
Visual</td><td>VICTORIES
0-0-0</td></tr>
<tr><td colspan="2">FUSES
.1 Nose, .025 Tail</td><td>FLAK
None</td></tr>
</table>

Mission	Description	ESCORT	WEATHER	LOSSES	BOMBS	BOMBING	VICTORIES	FUSES	FLAK
97 Belgrade, YUG - Sava East Railroad Bridge 5/9/1944	Proceeded to target area but found it 10/10 cloud covered and proceeded to the first alternate target where similar conditions were found. All bombs were therefore returned to base.	None	10/10	1-0-0	1000lb RDX	Did Not Bomb	0-0-0	.1 Nose, .025 Tail	S-I-H
98 Belgrade, YUG - Sava East Railroad Bridge6/9/1944	Target completely smoke obscured, first bombs fell South of target and walked into smoke with one direct hit on North approach.	14 FG	CAVU	0-0-1	1000lb RDX	Visual	0-0-0	.1 Nose, .025 Tail	I-A-H
99 Lyon, FRA - Bron Airdrome Supply Mission 10/9/1944	54 aircraft were dispatched from the 451BG and 461BG. 45 B-24s successfully accomplished the mission and return to base. 9 aircraft are unaccounted for, but believed to have landed at friendly fields.	None	N/A	1-0-0	N/A	N/A	0-0-0	N/A	N/A
100 Lyon, FRA - Bron Airdrome Supply Mission 12/9/1944	54 B-24s were dispatched from the 451, 461, and 484 Bomb Groups (49 Wing) 14 B-24s are believed to be at friendly fields.	None	N/A	0-0-0	N/A	N/A	0-0-0	N/A	N/A
101 Lyon, FRA - Bron Airdrome Supply Mission 13/9/1944	55 B-24s were dispatched without escort from the 451, 461, and 484 Bomb Groups (49 Wing). 54 B-24s successfully completed their mission. 1 B-24 returned early. 17 B-24s are at friendly fields.	None	N/A	0-0-0	N/A	N/A	0-0-0	N/A	N/A

102 Lyon, FRA - Bron Airdrome Supply Mission 15/9/1944

53 B-24s were dispatched without escort from the 451, 461, and 484 Bomb Groups (49 Wing) on a special supply mission to S. FRANCE. All A/C successfully delivered supplies and 49 B-24s returned to bases. The other 4 A/C are at friendly fields.

ESCORT	WEATHER	LOSSES
None	N/A	0-0-0
BOMBS	BOMBING	VICTORIES
N/A	N/A	0-0-0
FUSES		FLAK
N/A		N/A

103 Lyon, FRA - Bron Airdrome Supply Mission 16/9/1944

54 B-24s were dispatched without escort from the 451, 461, 484 Bomb Groups (49 Wing) on a special supply mission to Southern France. All aircraft successfully delivered supplies and 47 returned to base. 6 B-24s remained at French bases and I B-24 landed at a friendly field.

ESCORT	WEATHER	LOSSES
None	N/A	0-0-0
BOMBS	BOMBING	VICTORIES
N/A	N/A	0-0-0
FUSES		FLAK
N/A		N/A

104 Lyon, FRA - Bron Airdrome Supply Mission 17/9/1944

54 B-24s were dispatched without escort from the 451, 461, 484 Bomb Groups (49 Wing) on a special supply mission to Southern France. 8 B-24s successfully delivered supplies. Due to adverse weather 15 B-24s returned supply load to base. 26 B-24s landed loads at friendly fields and 5 are unaccounted for.

ESCORT	WEATHER	LOSSES
None	N/A	0-0-0
BOMBS	BOMBING	VICTORIES
N/A	N/A	0-0-0
FUSES		FLAK
N/A		N/A

105 Lyon, FRA - Bron Airdrome Supply Mission 22/9/1944

69 B-24s were dispatched without escort from the 451, 461, and 484 Bomb Groups (49 Wing). 68 B-24s successfully completed the mission, unloading at an A/D in Southern FRANCE. 1 B-24 returned early.

ESCORT	WEATHER	LOSSES
None	N/A	0-0-0
BOMBS	BOMBING	VICTORIES
N/A	N/A	0-0-0
FUSES		FLAK
N/A		N/A

106 Athens, GRC - Eleusis, Airdrome 24/9/1944	ESCORT 1FG	WEATHER CAVU	LOSSES 0-0-1
North dispersal area covered with heavy concentration of bombs. Taxi strip received many hits and a few were scored on North end of A/D. First Phase photo report shows one wing of probable administration building, West of hangers, received direct hit. Dispersal area only partially covered by photo.	BOMBS 100lb Clusters	BOMBING Visual	VICTORIES 0-0-90
	FUSES		FLAK S-I-H

107 Athens, GRC - Submarines in area 25/9/1944	ESCORT 14FG	WEATHER CAVU	LOSSES 1-0-3
24 N-24s of the 461 Bomb Group dropped 50.25 tons of bombs in two attacks from 21,000/22,000 feet on PIRAEUS Harbor installations. Bombs fell North of target in built up area and South of target in water. 3 small ships in harbor received several near misses.	BOMBS 500lb RDX	BOMBING Visual	VICTORIES 0-0-0
	FUSES .1 Nose, .025 Tail		FLAK M-I-H

Only one aircraft was lost on a combat mission during the month when it developed engine trouble and one other was lost (non-combat) when it burned to the ground on the runway at Lyon, France while delivering barrels of fuel. One man died in action and seventeen were wounded.

Twelve new crews joined the ranks of the 461st during September.

October 1944

Increasingly bad weather during October limited the group to only fifteen missions. Because the weather conditions were causing severe problems for large scale formation flying, the Air Force ordered smaller, two formation missions labeled "Red Force/Blue Force" or "Force Able/ Force Baker." October was also a time when flak batteries surrounding the remaining targets became more and more concentrated. In response Fifteenth aircraft, instead of bombing from 20,000, 21,000, and 22,000 feet to as high as 24,000, 25,000, and 26,000 feet. Field orders also increasingly called for the use of RDX bombs in place of the earlier general purpose bombs that had been used a great deal in the past.

108 Munich, DEU - West Marshalling Yard (Able) 4/10/1944	ESCORT	WEATHER	LOSSES
7 A/C were shot down and 16 were damaged by flak. In spite of the enemy resistance the group was rated superior. The bomb pattern traveled from the West to the center of the marshalling yard. Hits scored on rail lines, goods wagons, and trackage. Sadly, lost when his ship received a direct hit, Squadron Commander Maj. William Tallant, had just been promoted that day.	P-38	CAVU	7-1-13
	BOMBS	BOMBING	VICTORIES
	500lb RDX	Pathfinder	0-0-0
	FUSES		FLAK
	.1 Nose, No Delay Tail		I-A-H

109 Pinzano, ITA - South Railroad Bridge (Baker) 4/10/1944	ESCORT	WEATHER	LOSSES
Finding the target obscured by 7/10 cloud cover, the group was required to make five runs before dropped their bombs in a concentrated pattern. Due to the cloud cover the results were unobserved. However, the mission was rated superior.	None	7/10	0-0-0
	BOMBS	BOMBING	VICTORIES
	500lb RDX	Visual	0-0-0
	FUSES		FLAK
	.1 Nose, No Delay Tail		None

110 Vienna, AUT (Blue) Szombathely Marshalling Yard, HUN 7/10/1944	ESCORT	WEATHER CAVU	LOSSES 0-0-0
Late forming up due to bad weather at the assembly area, the group was late arriving at the rendezvous to find the other groups had left. Without fighter coverage it was decided to bomb the alternate. The Szombathely M/Y was selected. Damage included hits on the center choke point, railroad station, and overpass across the center choke point, turntable, and loco shop.	BOMBS 500lb RDX	BOMBING Visual	VICTORIES 0-0-0
	FUSES .1 Nose, No Delay Tail		FLAK None

111 Komarom, HUN (Red Force) - Marshalling Yard 7/10/1944	ESCORT None	WEATHER CAVU	LOSSES 0-0-0
The second half of the double mission was also late and chose to bomb an alternate target rather than face enemy fighters without an escort. The Marshalling Yard at KOMAROM was hit, scoring an impressive 91% even though eight aircraft had become separated from the group while attempting to climb through the solid deck of clouds, leaving only nine over target.	BOMBS 500lb GP	BOMBING Visual	VICTORIES 0-0-0
	FUSES .1 Nose, .025 Tail		FLAK S-I-H

112 Castelfranco Venito, ITA - Railroad Bridge 10/10/1944	ESCORT None	WEATHER 10/10	LOSSES 0-0-0
10/10 cloud cover made locating the target impossible. Some aircraft returned bombs to base while others jettisoned bombs in the Adriatic.	BOMBS 500lb RDX	BOMBING Did Not Bomb	VICTORIES 0-0-0
	FUSES .1 Nose, .025 Tail		FLAK S-I-H

113 Vienna, AUT - Vosendorf Oil Refinery 11/10/1944	ESCORT None	WEATHER 10/10	LOSSES 0-0-0
17 aircraft attempted to bomb VOSENDORF oil refinery, but 10/10 cloud cover turned the group back without bombing. 16 aircraft returned their bombs to base. 1 aircraft jettisoned their bombs in the KRANJ area.	BOMBS 500lb RDX	BOMBING Did Not Bomb	VICTORIES 0-0-0
	FUSES .1 Nose, .025 Tail		FLAK None

114 Bologna, ITA - Stores Depot 12/10/1944

Numerous hits on West half of stores depot with direct hits on storage warehouses and installations. Other hits on 2 warehouses in North Central portion of stores depot.

ESCORT	WEATHER	LOSSES
	CAVU	0-0-0
BOMBS	BOMBING	VICTORIES
100lb GP	Visual	0-0-0
FUSES		FLAK
		S-I-H

115 Vienna, AUT - Gas Works 13/10/1944

Bomb strike photos show target smoke covered by pots and previous attacks probably cut through line to KORNEUBURG. Most bombs fell just South of target. Direct hits were scored on assembly machine, repair, and workshops. One hit on North end of M/Y resulting in large explosions. Six direct hits on choke point.

ESCORT	WEATHER	LOSSES
1FG	8/10	2-1-16
BOMBS	BOMBING	VICTORIES
500lb RDX	Pathfinder	0-0-0
FUSES		FLAK
.1 Nose, .025 Tail		I-A-H

116 Bratislava, CZE - Marshalling Yard 14/10/1944

Assigned target was ODERTAL, but weather made bombing impossible. Target of opportunity was BRATISLAVA M/Y. The East end of the M/Y was hit as well as the North end of the small harbor. The South end of the Oil refinery was hit, but no explosion was seen. A direct hit on the motor bridge and several near misses were scored. Bombed visually because PFF failed to function.

ESCORT	WEATHER	LOSSES
14/332FG	7/10	0-0-4
BOMBS	BOMBING	VICTORIES
500lb RDX	Visual	0-0-0
FUSES		FLAK
.1 Nose, .025 Tail		M-A-H

117 Linz, AUS - Herman Goering Tank Work 16/10/1944

Bombed Herman Goering Tank Work Benzene Plant. Bombing done by Pathfinder due to 10/10 cloud cover. Results were unobserved.

ESCORT	WEATHER	LOSSES
82FG	10/10	1-0-1
BOMBS	BOMBING	VICTORIES
500lb RDX	Pathfinder	0-0-0
FUSES		FLAK
.1 Nose, .025 Tail		I-A-H

118 Vienna, AUS - Vosendorf Oil Refinery 17/10/1944

Adverse weather made bombing impossible. Bombs were returned to base.

ESCORT	WEATHER	LOSSES
82FG	10/10	0-0-0
BOMBS	BOMBING	VICTORIES
500lb RDX	Did Not Bomb	0-0-0
FUSES		FLAK
.1 Nose, .025 Tail		None

119 Milan, ITA - Isotta Fraschini Works 20/10/1944

Concentration of hits along Northwest edge of target with direct hits on three long assembly buildings. Barracks area North of target area received at several direct hits on at least 10 buildings.

ESCORT	WEATHER	LOSSES
None	CAVU	0-0-0
BOMBS	BOMBING	VICTORIES
500lb RDX	Visual	0-0-0
FUSES		FLAK
.1 Nose, .025 Tail		S-I-H

120 Munich DEU - Allach Bayerische Motoren Werke 23/10/1944

Due to cloud cover bombing was done by PFF with unobserved results.

ESCORT	WEATHER	LOSSES
1FG , 82FG	10/10	0-1-3
BOMBS	BOMBING	VICTORIES
500lb RDX	Pathfinder	0-0-0
FUSES		FLAK
.1 Nose, .025 Tail		I-A-H

121 Regensburg, DEU - Winterhafen Oil Storage 29/10/1944

Adverse weather forced group to return to base with bombs.

ESCORT	WEATHER	LOSSES
332FG, 31FG	10/10	0-0-0
BOMBS	BOMBING	VICTORIES
500lb/Incind	Pathfinder	0-0-0
FUSES		FLAK
.1 Nose, .025 Tail/Incendiary		None

122 Podgorics, YUG - Troop Concentrations31/10/1944

10/10 cloud cover forced group to abandon the target. Bombs were returned to the base.

ESCORT	WEATHER	LOSSES
None	10/10	0-0-0
BOMBS	BOMBING	VICTORIES
500lb RDX	Pathfinder	0-0-0
FUSES		FLAK
.1 Nose, .025 Tail		None

Over fifteen missions the group logged 2,703 combat hours. There was no enemy fighter opposition at all during the month. Aircraft losses included seven hit by flak, two involved in a mid-air collision, and one hit by bombs. Thirty two men were KIA, seventeen were WIA and thirty four became POWs.

Twenty six new crews arrived during the month.

November 1944

Bad weather continued to cause problems for large formations, which compelled the group to again fly two missions on a single day. Also, to employ a new tactic which responded to weather issues, pathfinder ships were employed in small numbers to bomb above 10/10 cloud cover, or were dispatched on night missions individually to bomb above a complete undercast.

Three new types of bombs were introduced during November. In addition to RDX, incendiary, and frags, the group used bombs with delayed fuses with six hour to 72 hour delays, bombs loaded with propaganda leaflets, and "screamers" fastened to bombs on two missions.

Over nineteen missions there were 421 aircraft airborne with 47 early returns. Accuracy for the month was rarely scored because of the limited number of missions the group bombed visually. The average score dropped from 63.4 during October to 31.4 in November dropping the group to 10th position in the Fifteenth Air Force.

123 Graz, AUS - Marshalling Yard 1/11/1944	ESCORT	WEATHER	LOSSES
Original target was VIENNA, Austria, however heavy cloud cover forced the group to choose an alternate. The best option would be LINZ, Austria. However, a heavy ground fog and the loss of the radar set caused the mission to be unsuccessful. Bombs were seen to hit in the vicinity of a small industrial area approximately 13,000 feet from the target.	305FW	2/10 Ext. Haze	0-0-7
	BOMBS	BOMBING	VICTORIES
	500lb RDX	Pathfinder	0-0-0
	FUSES		FLAK
	.1 Nose, .025 Tail		I-A-H

124 Klagenfurt, AUT - Marshalling Yard
3/11/1944

Four aircraft bombed the marshalling yard at KLAGENFURT by PFF. Results were unobserved due to 10/10 undercast, however the bombardiers believe they hit the target.

ESCORT	WEATHER	LOSSES
None	10/10	0-0-0
BOMBS	BOMBING	VICTORIES
500lb RDX	Pathfinder	0-0-0
FUSES		FLAK
.1 Nose, .01 Tail		None

125 Augsburg, DEU - Marshalling Yard
4/11/1944

The 461st put up 36 A/C against the marshalling yard at AUGSBURG. However, there was a 10/10 undercast at the target forcing the group to bomb by PFF with unobserved results.

ESCORT	WEATHER	LOSSES
82FG	10/10	0-0-2
BOMBS	BOMBING	VICTORIES
500lb RDX	Pathfinder	0-0-0
FUSES		FLAK
.1 Nose, .01 Tail		H-S-I

126 Vienna, AUT - Florisdorf Oil Refinery
5/11/1944

10/10 undercast forced the group to bomb by PFF with unobserved results. 461st gunners claimed one E/A shot down, one probable, and two more damaged.

ESCORT	WEATHER	LOSSES
305/306FW	10/10	0-0-1
BOMBS	BOMBING	VICTORIES
RDX + Incind.	Pathfinder	1-1-2
FUSES		FLAK
Two long delay per aircraft		I-I-H

127 Sarajevo, YUG - Ali Pasin Most Marshalling Yard 5/11/1944

A 6 aircraft formation was assigned to attack the Ali Pasin Most Marshalling Yard. However, they were forced to turn back from the target due to poor weather. A 360 degree circle was made in an attempt to find the target without success. Bombs were returned to base.

ESCORT	WEATHER	LOSSES
None	10/10	0-0-0
BOMBS	BOMBING	VICTORIES
500lb GP	Did Not Bomb	0-0-0
FUSES		FLAK
.1 Nose, .01 Tail		None

128 Bolzano, ITA - Power Sub Station
6/11/1944

Bolzano Power Station - Concentrated pattern of bombs in target area with three near missed on Transformer House. Highway and approach to highway bridge cut in 3 places. ORA Power Station - Two direct hits and several near misses on Power House. Double track RR line to BOLZANO cut North of siding by three direct hits.

ESCORT	WEATHER	LOSSES
P-38	HAZE	0-1-7
BOMBS	BOMBING	VICTORIES
500lb RDX	Visual	0-0-0
FUSES		FLAK
.1 Nose, .025 Tail		I-A-H

129 Vienna, AUT - South Ordnance Depot 6/11/1944			
22 aircraft attacked the VIENNA Ordnance Depot. Facing 10/10 cloud cover again the group was forced to bomb by PFF. The first flight hit the power station causing severe damage. The second flight missed the target completely, hitting an area on the river causing no damage.	ESCORT P-38	WEATHER 10/10	LOSSES 1-0-0
	BOMBS 500lb RDX	BOMBING Pathfinder	VICTORIES 0-0-0
	FUSES .1 Nose, .025 Tail		FLAK I-A-H

130 Sarajevo, YUG - Ali Pasin Most Marshalling Yard 7/11/1944			
All tracks cut or blocked. Approximately 200 units of rolling stock destroyed or damaged. One car shop destroyed and another virtually destroyed. Larger shop severely damaged and barracks destroyed.	ESCORT 325FG	WEATHER 7/10 W/ HAZE	LOSSES 0-0-16
	BOMBS 500lb RDX	BOMBING Visual	VICTORIES 0-0-0
	FUSES .1 Nose, .01 Tail		FLAK I-A-H

131 Sillien, AUT - Highway Bridge (Alternate) 11/11/1944			
Originally assigned to bomb LINZ, Austria, adverse weather forced the group to find a target of opportunity that could be bombed visually. Using 500lb RDX and 500lb incendiary bombs they attacked the highway bridge at SILLIEN, Austria through an undercast with unobserved results.	ESCORT 14FG	WEATHER 8/10	LOSSES 0-0-5
	BOMBS 500lb GP	BOMBING Visual	VICTORIES 0-0-0
	FUSES .1 Nose, .01 Tail		FLAK None

132 Innsbruck, AUT - Marshalling Yard 15/11/1944			
Four aircraft were assigned to bomb the INNSBRUCK M/Y using PFF. Three aircraft bombed through 10/10 undercast with unobserved results. The fourth aircraft was lost for undetermined reasons.	ESCORT None	WEATHER 10/10	LOSSES 1-0-0
	BOMBS 500lb RDX	BOMBING Pathfinder	VICTORIES 0-0-0
	FUSES .1 Nose, .01 Tail		FLAK None

133 Munich, DEU - Munich West Marshalling Yard 16/11/1944

28 aircraft, each loaded with 5 500lb RDX bombs, were assigned to bomb the M/Y at MUNICH. Unfortunately, the lead navigator accidentally hit a switch and released the bomb load. All the other aircraft released on the lead plane so the target was abandoned and the group returned to base.

ESCORT	WEATHER	LOSSES
P-51	10/10	0-0-0
BOMBS	BOMBING	VICTORIES
500lb RDX	Pathfinder	0-0-0
FUSES		FLAK
.1 Nose, .01 Tail		None

134 Vienna, AUT - Florisdorf Oil Refinery 17/11/1944

16 aircraft were assigned to bomb the Florisdorf Oil Refinery. Bombing was done by PFF over 10/10 undercast with unobserved results. All A/C returned to base.

ESCORT	WEATHER	LOSSES
P-38	10/10	0-0-0
BOMBS	BOMBING	VICTORIES
500lb RDX	Pathfinder	0-0-0
FUSES		FLAK
.1 Nose, .01 Tail		M-I-H

135 Blechhammer, DEU - Synthetic Oil Refinery 17/11/1944

23 aircraft took off for the BLECHHAMMER Synthetic Oil Refinery. There were three early returns. Bombing was done by PFF over 10/10 undercast with unobserved results. All A/C returned to the base except one which landed at Vis.

ESCORT	WEATHER	LOSSES
1FG/14FG	10/10	0-0-0
BOMBS	BOMBING	VICTORIES
500lb RDX	Pathfinder	0-0-0
FUSES		FLAK
.1 Nose, .01 Tail		M-I-H

136 Villafranca, ITA - Airdromes 18/11/1944

Bomb strike photos show North and West sides of A/D were well hit. Runways up to hangers received hits. While not ordered, P-51s from the 306BW, who were conducting a sweep in the area, provided fighter cover for the group.

ESCORT	WEATHER	LOSSES
None	CAVU W/ HAZE	0-0-2
BOMBS	BOMBING	VICTORIES
Frag Clusters	Visual	0-0-0
FUSES		FLAK
Instantaneous		S-I-H

137 Vienna, AUT - Vosendorf Oil Refinery 19/11/1944

Bombing was done by PFF through 10/10 undercast. One aircraft, flown by Lt. Farnham, was lost to flak.

ESCORT	WEATHER	LOSSES
1FG, 14FG	9/10	1-0-7
BOMBS	BOMBING	VICTORIES
500lb RDX	Pathfinder	0-0-0
FUSES		FLAK
.1 Nose, .01 Tail		I-A-H

138 Blechhammer, DEU - Synthetic Oil Refinery 20/11/1944	ESCORT	WEATHER	LOSSES
Direct hits on hydrogenation stalls, injector house and compressor house, with two explosions in the compressor house. Near misses on boiler house and storage tanks with fires resulting. Numerous strikes on other installations. North choke point at West side of refinery cut. Storage tracks on East side of refinery received several hits. 3 A/C ran out of fuel and were lost.	31FG, 332FG	CAVU	3-0-20
	BOMBS	BOMBING	VICTORIES
	500lb RDX	Visual	0-0-0
	FUSES		FLAK
	.1 Nose, .01 Tail		I-A-H

139 Novi Pazar, YUG - Troop Concentrations 21/11/1944	ESCORT	WEATHER	LOSSES
Concentration of bomb hits Southeast of NOVI PAZAR, South of intersection of highway leading Northward. Bomb concentration in fields. Bomb strike photos show hits across highway and RR tracks 3 miles East of CACAK.	None	4/10	0-0-9
	BOMBS	BOMBING	VICTORIES
	Frag Clusters	Visual	0-0-0
	FUSES		FLAK
	N/A		M-A-H

140 Munich, DEU - Marshalling Yard 22/11/1944	ESCORT	WEATHER	LOSSES
Undercast forced the group to bomb by PFF with unobserved results. Bombing was done with a four box diamond formation.	P-51s	10/10	0-0-6
	BOMBS	BOMBING	VICTORIES
	500lb RDX	Pathfinder	0-0-0
	FUSES		FLAK
	.1 Nose, .01 Tail		M-A-H

141 Munich, DEU - Marshalling Yard 25/11/1944	ESCORT	WEATHER	LOSSES
Three pathfinder aircraft participated in the group's first night mission. Two aircraft bombed the MUNICH M/Y by PFF. Oxygen leaks forced the third A/C to bomb Haiming, Austria.	None	9/10	0-0-0
	BOMBS	BOMBING	VICTORIES
	500lb RDX	Pathfinder	0-0-0
	FUSES		FLAK
	.1 Nose, .01 Tail		I-I-H

The group lost five aircraft to combat and one weather related loss. Twenty men were killed and other seventeen wounded. The group also destroyed one enemy fighter, probably one other, and damaged four more. Twenty new crews arrived during November.

December 1944

Seventeen combat missions were flown during December, seven against synthetic fuel refineries in Austria, Germany, and Czechoslovakia; eight against marshalling yards; and two more against railroad viaducts. It would prove to be the toughest month since July.

Tragedy would strike the group when a ferry aircraft bringing back personnel from R&R crashed into a small knoll trying to land in bad weather. Ten of the fifteen people on the plane were killed and the others seriously injured.

Mission	Description	ESCORT	WEATHER	LOSSES	BOMBS	BOMBING	VICTORIES	FUSES	FLAK
142 Blechhammer, DEU - Synthetic Oil Refinery 2/12/1944	Scattered craters throughout the plant with only light damage inflicted destroying three buildings, a warehouse, damaging others, and cutting sidings in eight places.	P-51s	1/10 HAZE	0-0-15	500lb RDX	Pathfinder	0-0-0	.1 Nose, .01 Tail	M-A-H
143 Innsbruck, AUT - Marshalling Yard 3/12/1944	Two aircraft made pathfinder attacks on the INNSBRUCK M/Y. The cloud cover prevented any observation or assessment of the mission. One aircraft returned safely to the base, but the other was compelled to land at a friendly base.	None	10/10	0-0-0	500lb RDX	Pathfinder	0-0-0	.1 Nose, .01 Tail	None
144 Maribor, YUG - Marshalling Yard 6/12/1944	Photos show bombs fell Northeast of target at West edge of railway repair shop and small sidings. Three direct hits and two near misses cut single track rail line to DRAVOGRAD. Car repair shop received one direct hit.	1 FG	2/10	0-0-9	500lb RDX	Visual	0-0-0	.1 Nose, .01 Tail	S-A-H

145 Innsbruck, AUT - Main Railroad Station (Night) 7/12/1944

Two aircraft were dispatched to make night Lone Wolf attacks on the main railroad tracks at INNSBRUCK, Austria. One ship was required to turn around when an engine caught fire. The other successfully bombed by PFF with unobserved results. German night fighters were active against American bombers from different groups, but were unsuccessful.

ESCORT	WEATHER	LOSSES
None	2/10	0-0-0
BOMBS	**BOMBING**	**VICTORIES**
500lb RDX	Pathfinder	0-0-0
FUSES		**FLAK**
.1 Nose, .01 Tail		M-I-H

146 Moosbierbaum, AUT - Oil Refinery (Night) 8/12/1944

Another two-plane Lone Wolf mission, this one ending without bombing the target. One ship encountered clear weather and abandoned the target to return to base. The other ship was attacked by five German fighters, but was able to fight off the attack and find cloud cover, returning to base.

ESCORT	WEATHER	LOSSES
None	CLEAR	0-0-0
BOMBS	**BOMBING**	**VICTORIES**
500lb RDX	Did Not Bomb	0-0-0
FUSES		**FLAK**
.1 Nose, .01 Tail		None

147 Vienna, AUT - Maitzlendorf Goods Station 11/12/1944

Photos show bombs fell South of the target. One direct hit on the rail line. Most bombs fell in a built up industrial area.

ESCORT	WEATHER	LOSSES
52FG	3/10	1-1-16
BOMBS	**BOMBING**	**VICTORIES**
500lb RDX	Pathfinder	0-0-0
FUSES		**FLAK**
.1 Nose, .01 Tail		I-A-H

148 Blechhammer, DEU - Synthetic Oil Refinery 12/12/1944

Six aircraft were assigned to attack BLECHHAMMER in three groups of two ships. However, One A/C failed to take off. Two were early returns, two others hit alternate targets one hit the assigned target. The ship that did hit the primary was forced to land at the ISLE OF VIS while on the way home to refuel.

ESCORT	WEATHER	LOSSES
None	10/10	0-2-0
BOMBS	**BOMBING**	**VICTORIES**
500lb RDX	Pathfinder	1-0-0
FUSES		**FLAK**
.1 Nose, .025 Tail		M-I-H

149 Linz, AUT - Marshalling Yard, Rolling Stock 15/12/1944

Bombing was done by PFF and results are unknown. One aircraft developed engine trouble and the crew was forced to abandon the ship over Yugoslavia. Lt. Marshall and several of his crew were able to reach friendly lines, but three men landed in the Sava River and drowned.

ESCORT	WEATHER	LOSSES
14 FG	10/10	1-0-0
BOMBS	BOMBING	VICTORIES
500lb RDX	Pathfinder	0-0-0
FUSES		FLAK
.1 Nose, .01 Tail		M-I-H

150 Brux, CZE - Oil Refinery 16/12/1944

The 461st bombed the BRUX Oil Refinery through complete cloud cover by PFF. Results were unknown. One aircraft was lost to flak.

ESCORT	WEATHER	LOSSES
31FG	10/10	1-0-0
BOMBS	BOMBING	VICTORIES
500lb RDX	Pathfinder	0-0-0
FUSES		FLAK
.1 Nose, .01 Tail		I-I-H

151 Odertal, DEU - Oil Refinery 17/12/1944

Bombing was done through an undercast by PFF with unobserved results. The crews had been briefed not to lower their ball turrets until they reached the IP to conserve fuel. Prior to the IP the group was attacked by 50 German fighters who, seeing the ball turrets were still raised, came in low firing rockets into the bomb bays. Ten aircraft were lost as a result.

ESCORT	WEATHER	LOSSES
P-38	10/10	10-0-4
BOMBS	BOMBING	VICTORIES
500lb RDX	Pathfinder	19-5-0
FUSES		FLAK
.1 Nose, .01 Tail		S-I-H

152 Blechhammer, DEU - Synthetic Oil Refinery 18/12/1944

Complete cloud cover once again required bombing by pathfinder. Results were unobserved. One aircraft was lost.

ESCORT	WEATHER	LOSSES
1FG, 52FG	10/10	1-0-0
BOMBS	BOMBING	VICTORIES
500lb RDX	Pathfinder	0-0-0
FUSES		FLAK
.1 Nose, .025 Tail		I-A-H

153 Blechhammer, DEU - Synthetic Oil Refinery 19/12/1944	ESCORT 82,114,325FG	WEATHER 10/10	LOSSES 0-0-6
Six 461st B-24s, having become lost in a snow storm, bombed a target of opportunity near STERNBERG by PFF. Results were undefined.	BOMBS 500lb RDX	BOMBING Pathfinder	VICTORIES 0-0-0
	FUSES .1 Nose, .01 Tail		FLAK M-I-H

154 Villach, AUT - Marshalling Yard 20/12/1944	ESCORT 1, 14, 325FG	WEATHER 10/10	LOSSES 0-0-2
A two flight formation was assigned to bomb the synthetic Oil Refinery at BRUX, Czechoslovakia. The two flights became separated in bad weather. As a result one flight bombed VILLACH, Austria and the other bombed LINZ, Austria by PFF with undetermined results.	BOMBS 500lb RDX	BOMBING Pathfinder	VICTORIES 0-0-0
	FUSES .1 Nose, .025 Tail		FLAK S-I-H

155 Wels, AUT - Marshalling Yard 25/12/1944	ESCORT 1FG, 52FG	WEATHER CAVU W/ HAZE	LOSSES 0-0-0
The primary, BRUX was not attacked due to bad weather at the base, which caused a fuel shortage necessitating alternate targets. Wels M/Y hit with good concentration, scoring hits on wagons and trackage. Double line to LINZ and central choke point hit. West of M/Y covered by hits to goods wagons and trackage. Line to LAMBOCH and HAIDING were probably cut by direct hit.	BOMBS 500lb RDX	BOMBING Visual	VICTORIES 0-0-0
	FUSES .1 Nose, .025 Tail		FLAK I-A-H (Brux)

156 Venzone, ITA - Viaduct 27/12/1944	ESCORT 1FG	WEATHER CAVU	LOSSES 0-0-0
Majority of bombs hit West of viaduct with 6 probable hits and several near misses. Highway East of viaduct cut by three hits. Six 461st ships bombed VILLACH M/Y visually as an alternate because the primary was obscured by smoke from previous bombing. Bombs fell North of target on the edge of built up area. Two B-24s returned early.	BOMBS 1000lb RDX	BOMBING Visual	VICTORIES 0-0-0
	FUSES .1 Nose, .025 Tail		FLAK None

157 Venzone, ITA - Viaduct 28/12/1944	ESCORT None	WEATHER CAVU	LOSSES 0-0-0
Target was attacked visually by all three groups of the 49th Bomb Wing. (451st, 461st, & 484th). The 76 ships of the 49th Wing dropped 218 tons of RDX bombs from 17,500/20,000 feet. Two spans at two separate points were destroyed and the road was damaged by two direct hits. 2 FW 190s were seen in the area and 7-8 other a/c followed the formation to the coast.	BOMBS 1000lb RDX	BOMBING Visual	VICTORIES 0-0-0
	FUSES .1 Nose, .025 Tail		FLAK None

158 Rosenheim, DEU - Marshalling Yard 29/12/1944	ESCORT 31 FG	WEATHER CAVU	LOSSES 0-0-0
13 aircraft dropped 25 tons of RDX bombs on CASTELFRANCO VENITO Locomotive Depot, an alternate target, using visual method. Photos show a close pattern fell on the loco depot and storage yard. 6 other ships hit ROSENHEIM by PFF bombing with unobserved results. On return, three flights let down through a hole in the land at emergency fields.	BOMBS 1000lb RDX	BOMBING Pathfinder	VICTORIES 0-0-0
	FUSES .1 Nose, .025 Tail		FLAK None

December proved to be one of the worst months for losses of the entire war. Including one non-combat loss of a ferry aircraft, there were fifteen ships lost. Ten of those losses occurred on 17 December when over 100 German fighters attacked the group, coming in low and firing rockets up into the bomb bays. Personnel losses included 33 killed in action, 28 wounded in action, and 83 prisoners of war.

January 1945

January brought with it the worst weather of the war. It was a month of rain, snow, high winds, and deep mud, which caused multiple stand-downs. The result was eight missions and fifteen cancellations.

Twenty four new crews arrived resulting in too many crews and not enough aircraft.

Mission			
159 Trento, ITA - North Marshalling Yard 4/1/1945	ESCORT 31FG	WEATHER CAVU	LOSSES 0-0-3
Bomb strike photos show pattern of bombs falling across target with hits along center sidings and South choke point. Near misses scored on the station, 30 percent of sidings blocked. All stock in goods yard blocked until repairs are effected.	BOMBS 500lb RDX	BOMBING Visual	VICTORIES 0-0-0
	FUSES .1 Nose, .025 Tail		FLAK M-I-H
160 Zagreb, YUG - Marshalling Yard - East Side 5/1/1945	ESCORT 325FG	WEATHER 9/10	LOSSES 0-0-0
All three groups of the 49th BW were dispatched to bomb the marshalling yard at ZAGREB. As always with targets in Yugoslavia, they were ordered to bomb visually only. When they arrived they found 9/10 cloud cover over target. The group returned their bombs to the base.	BOMBS 500lb RDX	BOMBING Did Not Bomb	VICTORIES 0-0-0
	FUSES .1 Nose, .025 Tail		FLAK M-I-H
161 Klagenfurt, AUT - Marshalling Yard 8/1/1945	ESCORT 332FG	WEATHER 10/10	LOSSES 1-0-3
The group was tasked with bombing the marshalling yard at LINZ, Austria, but a high cloud cover made bombing impossible. They were finally able to bomb their second alternate target, the KLAGENFURT marshalling yard. Bombing was done by pathfinder with unobserved results. One aircraft was lost due to mechanical difficulties and weather during the mission.	BOMBS 500lb RDX	BOMBING Visual	VICTORIES 0-0-0
	FUSES .1 Nose, .01 Tail		FLAK None

162 Treviso, ITA - Marshalling Yard 15/1/1945			
Unable to bomb the primary target because of an instrument problem the group attacked the alternative, TREVISO marshalling yard visually. Photos show bombs straddling the M/Y, but smoke and dust prevent accurate assessment. Three hits are visible on barracks 1 mile Southeast.	ESCORT 325 FG	WEATHER CAVU	LOSSES 0-0-8
	BOMBS 500lb RDX	BOMBING Visual	VICTORIES 0-0-0
	FUSES .1 Nose, .025 Tail		FLAK M-A-H

163 Brod, YUG - Sava River Railroad Bridge 19/1/1945			
North end of bridge covered by excellent concentration of bombs with several hits and near misses. One direct hit on South end of bridge. Hits are also visible on an old fort North of the bridge, where direct hits were scored on warehouse type buildings. The rail line to ZAGREB was hit Southeast of the target.	ESCORT None	WEATHER CAVU	LOSSES 0-0-19
	BOMBS 500lb RDX	BOMBING Visual	VICTORIES 0-0-0
	FUSES .1 Nose, .025 Tail		FLAK H-S-A

164 Linz, AUD - North Main Marshalling Yard 20/1/1945			
Concentrated pattern of bombs fell on North choke point of M/Y and on the RR junction. Freight lines and rail yards leading North of freight yards received a number of direct hits to goods wagons and cut track. One string of bombs fell across loco depot scoring several direct hits on repair facilities.	ESCORT 52 FG	WEATHER 4/10	LOSSES 2-1-20
	BOMBS 100lb Clusters	BOMBING Visual	VICTORIES 0-0-0
	FUSES .1 Nose, .025 Tail		FLAK I-A-H

165 Moosbierbaum, AUT - Oil Refinery (Red Force) 31/1/1945			
Red Force had lined up for their attack on MOOSBIERBAUM, but was forced off the run by another group. A second attempt to line up failed so the group was forced to return their bombs to the base. One aircraft ran out of fuel and was forced to ditch off the coast of Yugoslavia. (Six KIA, 4 rescued).	ESCORT 325FG, 332FG	WEATHER 10/10	LOSSES 2-0-0
	BOMBS 500lb RDX	BOMBING Did Not Bomb	VICTORIES 0-0-0
	FUSES .1 Nose, .025 Tail		FLAK S-A-H

166 Moosbierbaum, AUT - Oil Refinery (Blue Force) 31/1/1945	ESCORT 325FG, 332FG	WEATHER 10/10	LOSSES 0-0-0
The Blue Force successfully attacked MOOSBIERBAUM using PFF. Due to 10/10 cloud cover no results could be determined.	BOMBS 500lb RDX	BOMBING Pathfinder	VICTORIES 0-0-0
	FUSES .1 Nose, .025 Tail		FLAK S-I-H

Two aircraft were lost to flak, two ran out of fuel on the way home, and a fifth loss was caused by weather. Twenty six men were KIA. thirteen were WIA, and 24 became POWs.

February 1945

Weather began to improve during the month. While six of fifteen missions were cancelled in the first half of the month, there were no cancellations in the second half for a total of twenty-one completed missions.

Visual bombing turned in a score of 27.3 to rank the 461st fourteenth in the Air Force. A bright spot was operational efficiency. Of 530 aircraft airborne on combat missions, there were only 28 early returns.

167 Graz, AUT - Marshalling Yard 1/2/1945	ESCORT 332FG	WEATHER 10/10	LOSSES 0-0-0
Five bomb wings of the Fifteenth Air Force were assigned to bomb the marshalling yard at MOOSBIERBAUM, but only four groups were able to bomb the primary. The rest of the formation were assigned alternate targets. The 461st bombed the marshalling yard at GRAZ, Austria by PFF. Complete cloud cover made results impossible to determine.	BOMBS 500lb RDX	BOMBING Pathfinder	VICTORIES 0-0-0
	FUSES .1 Nose, .025 Tail		FLAK S-I-H

168 Straubing, DEU - Town Proper 5/2/1945	ESCORT 31FG	WEATHER 10/10	LOSSES 0-0-0
The group was briefed to bomb the REGENSBURG Winter Harbor Oil Storage Facility, Germany. They bombed through solid undercast with unobserved results. It was later determined that they had bombed 20 miles East of course and hit STRAUBING, the IP.	BOMBS 250lb Clusters	BOMBING Pathfinder	VICTORIES 0-0-0
	FUSES .1 Nose, .025 Tail		FLAK None
169 Vienna, AUT - Florisdorf Oil Refinery 7/2/1945	ESCORT 82FG	WEATHER CAVU	LOSSES 0-2-1
The group bombed using a combination of visual and offset methods. Bomb strike photos show strikes entering the target area from just outside smoke covered area. Other strikes are visible in the M/Y Northeast of FLORISDORF M/Y with scattered hits Southwest of Nordweptbahn bridge. Hits were observed in minor industries in North part of VIENNA.	BOMBS 500lb RDX	BOMBING Visual	VICTORIES 0-0-0
	FUSES .1 Nose, .025 Tail		FLAK I-I-H
170 Vienna, AUT - Florisdorf Oil Refinery 7/2/1945	ESCORT 82FG	WEATHER CAVU	LOSSES 0-1-3
The group was assigned to bomb the Florisdorf Oil Refinery for the second time in the same day. Bombs fell Southwest of the target in the VIENNA freight Yards. Flak was heavy, damaging 6 of the 16 aircraft and wounding 3 men.	BOMBS 500lb RDX	BOMBING Visual	VICTORIES 0-0-0
	FUSES .1 Nose, .025 Tail		FLAK I-A-H
171 Vienna, AUT - Central Repair Shops 8/2/1945	ESCORT 325FG	WEATHER 8/10	LOSSES 0-0-9
Every group in the Fifteenth Air Force bombed VIENNA except the 99th. The 461st bombed by PFF with unobserved results. As usual flak was heavy in VIENNA, hitting nine 461st ships.	BOMBS 500lb RDX	BOMBING Pathfinder	VICTORIES 0-0-0
	FUSES .1 Nose, .025 Tail		FLAK I-A-H

172 Moosbierbaum, AUT - Oil Refinery 9/2/1945			
Three Lone Wolf aircraft were assigned to attack the MOOSBIERBAUM Oil Refinery. As is required for Lone Wolf missions there was a total undercast at the target. They were opposed by flak batteries, but were not hit.	ESCORT None	WEATHER 10/10	LOSSES 0-0-0
	BOMBS 500lb RDX	BOMBING 3 PFF Only	VICTORIES 0-0-0
	FUSES .1 Nose, .025 Tail		FLAK I-I-H

173 Vienna, AUT - Central Repair Shops (Red Force) 13/2/1945			
Seventeen aircraft were assigned to attack the VIENNA Central Repair Shops. Weather was CAVU, but smoke pots completely obscured the target, which was completely missed by the group. One ship was lost to flak and 14 of the remaining 16 ships were damaged.	ESCORT 332FG	WEATHER CAVU	LOSSES 2-1-13
	BOMBS 500lb RDX	BOMBING Visual	VICTORIES 0-0-0
	FUSES .1 Nose, .025 Tail		FLAK I-A-H

173 Maribor, YUG - Central Repair Shops (Blue Force) 13/2/1945			
The eighteen ship Blue Force attacked the VIENNA Central Repair Shops in CAVU weather. Bombing was done visually with a 28.3% score. Sixteen aircraft were holed by flak. One of these ships crashed at the base killing Lt. Donald Williams. Lt. Williams was a replacement navigator on his first mission with the crew.	ESCORT 332FG	WEATHER CAVU	LOSSES 1-0-8
	BOMBS 500lb RDX	BOMBING Visual	VICTORIES 0-0-0
	FUSES .1 Nose, .025 Tail		FLAK M-A-H

174 Moosbierbaum, AUT - Oil Refinery 14/2/1945			
Red Force bombed the MOOSBIERBAUM Oil Refinery visually in spite of scattered clouds and smoke pots. Results showed only 7.1% of bombs hit within 1000 feet of the aiming point. The Blue Force bombed the same target using PFF with unobserved results.	ESCORT 31FG	WEATHER 2/10CAVU	LOSSES 0-0-2
	BOMBS 500lb RDX	BOMBING Visual/PFF	VICTORIES 0-0-0
	FUSES .1 Nose, .025 Tail		FLAK M-I-H

175 Vienna, AUT - Penzinger Marshalling Yard 15/2/1945

Red Force and Blue Force both attacked the Penzinger M/Y by PFF through a complete cloud cover. Results were unobserved due to the cloud cover. One crew member was wounded by flak.

ESCORT	WEATHER	LOSSES
332FG	2/10	0-0-0
BOMBS	BOMBING	VICTORIES
500lb RDX	Pathfinder	0-0-0
FUSES		FLAK
.1 Nose, .025 Tail		I-A-H

176 Rosenheim, DEU - West Marshalling Yard 16/2/1945

The target for the day was the ROSENHEIM M/Y. The group bombed visually due to the clearing weather. A close pattern of bombs fell in the East portion of the marshalling yard scoring hits among rolling stock and trackage. Three direct hits were scored on the through lines.

ESCORT	WEATHER	LOSSES
325FG, 82FG	CAVU	0-0-0
BOMBS	BOMBING	VICTORIES
500lb RDX	Visual	0-0-0
FUSES		FLAK
.1 Nose, .025 Tail		None

177 Trieste, ITA - Shipyards 17/2/1945

Primary target on this date had been the tank works at St Valentin, Italy, but a delay taking off due to weather conditions over the Alps forced the group to bomb an alternate target. The weight of the attack fell across the Marco Shipyards. Smoke from fires in the shipyard and industrial areas made further assessment impossible except for one hit on a warship in the harbor.

ESCORT	WEATHER	LOSSES
325FG	CAVU	0-0-1
BOMBS	BOMBING	VICTORIES
500lb RDX	Visual	0-0-0
FUSES		FLAK
.1 Nose, .025 Tail		M-I-H

178 St Valentin Tank Works, Austria 18/2/1945

The 461st took off to bomb the St Valentin Tank Works, but ran into heavy cloud cover over the Adriatic from 8,000 to 30,000 feet. Unable to penetrate the cloud cover, the group was forced to return to base with their bombs. There was no credit for the mission.

ESCORT	WEATHER	LOSSES
None	CAVU	0-0-0
BOMBS	BOMBING	VICTORIES
500lb RDX	Did not bomb	0-0-0
FUSES		FLAK
.1 Nose, .025 Tail		M-I-H

<table>
<tr><td colspan="2">179 Vienna Railroad Targets 20/2/1945</td><td>ESCORT
325FG</td><td>WEATHER
1/10</td><td>LOSSES
0-0-0</td></tr>
<tr><td colspan="2" rowspan="3">Fourteen group liberators attacked the South Station area using a combination of visual and PFF bombing. Photos show a good concentration of hits with 23 bombs falling on goods wagons. Three direct hits on warehouses. The remaining ships bombed the Fiume Shipyards visually. Photos show the group completely missed the target.</td><td>BOMBS
500lb RDX</td><td>BOMBING
Visual</td><td>VICTORIES
0-0-0</td></tr>
<tr><td colspan="2">FUSES
.1 Nose, .025 Tail</td><td>FLAK
S-I-H</td></tr>
</table>

<table>
<tr><td colspan="2">180 Vienna, AUT - South Station Area 21/2/1945</td><td>ESCORT
82FG</td><td>WEATHER
Haze</td><td>LOSSES
3-0-7</td></tr>
<tr><td colspan="2" rowspan="2">The first section bombed the South Station area, but they missed the target entirely. Flak was heavy, costing the group three ships with four more damaged. The second section dropped on the Weiner Neustadt aircraft engine factory with unknown results.</td><td>BOMBS
500lb RDX</td><td>BOMBING
Pathfinder</td><td>VICTORIES
0-0-0</td></tr>
<tr><td colspan="2">FUSES
.1 Nose, .025 Tail</td><td>FLAK
I-A-H</td></tr>
</table>

<table>
<tr><td colspan="2">181 Kempten, DEU - Marshalling Yard 22/2/1945</td><td>ESCORT
31FG</td><td>WEATHER
CAVU</td><td>LOSSES
0-1-5</td></tr>
<tr><td colspan="2" rowspan="2">The marshalling yard at KEMPTON, Germany was a target of opportunity. The main concentration fell on the South of the M/Y with four direct hits on the choke point. One large building was heavily damaged by at least two direct hits while other hits fell on car shops and a warehouse. Through tracks were probably cut. No credit was given because the target was not a briefed one.</td><td>BOMBS
500lb RDX</td><td>BOMBING
Visual</td><td>VICTORIES
0-0-0</td></tr>
<tr><td colspan="2">FUSES
.1 Nose, .025 Tail</td><td>FLAK
None</td></tr>
</table>

<table>
<tr><td colspan="2">182 Knittelfeld, AUT - Marshalling Yard 23/2 1945</td><td>ESCORT
52FG</td><td>WEATHER
4/10</td><td>LOSSES
0-0-0</td></tr>
<tr><td colspan="2" rowspan="2">The marshalling yard at KNITTELFELD was bombed as a target of opportunity. The group dropped their bombs after three runs on the target, scoring an impressive 49%. The roundhouse and through tracks were well covered.</td><td>BOMBS
500lb RDX</td><td>BOMBING
Visual</td><td>VICTORIES
0-0-0</td></tr>
<tr><td colspan="2">FUSES
.1 Nose, .025 Tail</td><td>FLAK
None</td></tr>
</table>

183 Linz AUT - S. Main Marshalling Yard 25/2/1945

The group was forced to bomb by PFF in spite of CAVU weather because of heavy smoke from previous bomb runs. Results were also unobserved due to the smoke.

ESCORT	WEATHER	LOSSES
325FG	10/10	0-0-0
BOMBS	BOMBING	VICTORIES
1000lb RDX	Pathfinder	0-0-0
FUSES		FLAK
.1 Nose, .025 Tail		I-A-H

184 Augsburg, DEU - Marshalling Yard 27/2/1945

The group dropped 1000lb RDX bombs by synchronous PFF with unobserved results. Heavy flak over the target resulted in thirteen aircraft out of twenty six being damaged. However, there were no losses.

ESCORT	WEATHER	LOSSES
332FG	5/10	0-0-13
BOMBS	BOMBING	VICTORIES
1000lb RDX	Pathfinder	0-0-0
FUSES		FLAK
.1 Nose, .025 Tail		I-A-H

185 Ora, ITA - Marshalling Yard 28/2/1945

This was a double header mission with both forces attacking the marshalling yard at ORA, Italy. The Red Force dropped visually with a score of 15.6, The Blue Force dropped with a score of 28.1. Photos show hits on the North and South ends of the yard with both choke points cut. Direct hits and near misses were scored on the station and among goods wagons and trackage.

ESCORT	WEATHER	LOSSES
31FG, 325FG	CAVU	0-0-9
BOMBS	BOMBING	VICTORIES
500lb RDX	Visual	0-0-0
FUSES		FLAK
.1 Nose, .025 Tail		M-I-H

Six aircraft were lost during the month. There were 15 crew members killed in action and 22 wounded. Eleven became prisoners of war.

March 1945

The group flew twenty-one combat missions during March. Of the 631 ships on the missions reached their briefed targets. They dropped 1,010 tons of the total 1,117 tons on targets, with an average bombing accuracy of 42.7. This score ranked thirteenth of the twenty-one groups in the Fifteenth Air Force. They also had the smallest percentage of early returns and non-effective sorties in the Air Force, with only 4.4 sorties in every one hundred listed as non-effective.

186 Moosbierbaum, AUT - Oil Refinery 1/3/1945	ESCORT 52FG	WEATHER 3/10	LOSSES 0-0-7
Another double header, the Blue Force was unable to find a way around bad weather in the LAKE BALATON area and was forced to return their bombs to the base. The Red Force finally found a hole in the weather and dropped their bombs visually. Bomb strike photos showed extensive damage to the target.	BOMBS 500lb RDX	BOMBING Pathfinder	VICTORIES 0-0-0
	FUSES .1 Nose, .025 Tail		FLAK M-A-H

187 Linz, AUT - North Marshalling Yard 2/3/1945	ESCORT 325FG	WEATHER 8/10	LOSSES 0-1-11
The group was forced to bomb by synchronous PFF. Cloud cover made any photo interpretation impossible.	BOMBS 500lb RDX	BOMBING Pathfinder	VICTORIES 0-0-0
	FUSES .1 Nose, .025 Tail		FLAK M-A-H

188 Graz, AUT - Main Marshalling Yard 4/3/1945	ESCORT 332FG	WEATHER 4/10	LOSSES 0-0-0
The mission of 4 March against GRAZ Main Marshalling Yard, Austria found Major Rider leading a large formation. The bombs fell across the target area with twenty-five strikes on cars in the yard and other rail installations. An overhead pass was hit, while still other bombs fell on. The tracks between the main marshalling yard and the freight yard. The mission was scored 26.7 percent.	BOMBS 500lb RDX	BOMBING Visual	VICTORIES 0-0-0
	FUSES .1 Nose, .025 Tail		FLAK S-I-H

189 Hegyeshalom, HUN - Marshalling Yard 8/3/1945	ESCORT P-51	WEATHER 4/10	LOSSES 0-0-0
Very heavy concentration of hits. All tracks cut on many places. Buildings, installations, and many goods wagons were destroyed or damaged in many places.	BOMBS 500lb RDX	BOMBING Visual	VICTORIES 0-0-0
	FUSES .1 Nose, .025 Tail		FLAK None

190 Graz, AUT (Blue Force) Marshalling Yard 9/3/1945	ESCORT P-38, P-51	WEATHER 4/10	LOSSES 2-2-4
Weather forced the Blue Force to abandon the primary target. The GRAZ M/Y, Austria and choose alternate GRAZ M/Y. Photos show numerous hits on railroad coach and freight car plants, warehouse, and freight yard at Southeast corner of the yard.	BOMBS 500lb RDX	BOMBING Pathfinder	VICTORIES 0-0-0
	FUSES .1 Nose, .025 Tail		FLAK M-A-H

190 Graz, AUT (Red Force) Marshalling Yard 9/3/1945	ESCORT 305/306BG	WEATHER 8/10	LOSSES 0-0-4
Red Force was also unable to bomb the primary and choose GRAZ M/Y as the alternate target. They also bombed by PFF with the same results as Blue Force.	BOMBS 500lb RDX	BOMBING Pathfinder	VICTORIES 0-0-0
	FUSES .1 Nose, .025 Tail		FLAK S-A-H

191 Vienna, AUT - Florisdorf Oil Refinery 12/3/1945	ESCORT 52FG	WEATHER 10/10	LOSSES 0-0-0
A 10/10 cloud cover forced the group to bomb by PFF. The cloud cover prevented observation of the results or assessment of the bomb strike photos. However, most crews reported good PFF runs. They also experienced bombing VIENNA without a single aircraft suffering flak damage.	BOMBS 500lb RDX	BOMBING Pathfinder	VICTORIES 0-0-0
	FUSES .1 Nose, .025 Tail		FLAK M-I-H

192 Regensburg, DEU - Main Marshalling Yard 13/3/1945	ESCORT 14FG	WEATHER 9/10	LOSSES 0-0-0
Once more the group was forced to bomb through complete cloud cover with unknown results.	BOMBS 500lb RDX	BOMBING Pathfinder	VICTORIES 0-0-0
	FUSES .1 Nose, .025 Tail		FLAK None

193 Weiner Neustadt, AUT - Marshalling Yard 14/3/1945	ESCORT P-38	WEATHER 10/10	LOSSES 0-0-0
The group was briefed for a visual mission against the marshalling yard and BRUCK, Austria, but a 10/10 cloud cover forced them to choose an alternate PFF target. WEINER NEUSTADT was bombed by PFF without any observed results.	BOMBS 500lb RDX	BOMBING Pathfinder	VICTORIES 0-0-0
	FUSES .1 Nose, .01 Tail		FLAK None

194 Weiner Neustadt, AUT - Marshalling Yard 15/3/1945

Today's target was to be a visual attack against Schwechat Oil Refinery at VIENNA, but again heavy cloud cover forced an alternate PFF mission. Again WIENER NEUSTADT was selected and bombed by PFF without any observed results.

ESCORT	WEATHER	LOSSES
82FG	9/10	1-0-0
BOMBS	**BOMBING**	**VICTORIES**
500lb RDX	Pathfinder	0-0-0
FUSES		**FLAK**
.1 Nose, .025 Tail		None

195 Moosbierbaum, AUT - Oil Refinery 16/3/1945

A clear day allowed the group to bomb the MOOSBIERBAUM Oil Refinery visually with excellent results. Bomb strike photos showed a concentration of hits in the North end of the refinery with many direct hits on the distillation units, liquid air compressors, and oil storage tanks.

ESCORT	WEATHER	LOSSES
325FG	CAVU	0-1-1
BOMBS	**BOMBING**	**VICTORIES**
500lb RDX	Visual	0-0-0
FUSES		**FLAK**
.1 Nose, .025 Tail		I-A-H

196 Muhldorf, DEU - Marshalling Yard 19/3/45

Photos show strikes covering the majority of the yard, particularly the center and west. Many direct hits on rolling stock, through lines and sidings tracks. Other facilities were hit including the station buildings, turntable, and coal piles. The east choke point received five direct hits. Parts of the target were obscured by heavy bomb smoke.

ESCORT	WEATHER	LOSSES
332FG	CAVU	0-0-1
BOMBS	**BOMBING**	**VICTORIES**
100lb GP	Visual	0-0-0
FUSES		**FLAK**
.1 Nose, .025 Tail		S-A-H

197 Wels, AUT - Main Marshalling Yard 20/3 1945

Photos show a good concentration of hits covering the main yard, loco depot and holding sidings. Many hits on roundhouse, through lines, tracks, repair shops and buildings adjacent to the depot. Many fires and explosions were visible throughout the yard.

ESCORT	WEATHER	LOSSES
52FG	CAVU	0-0-0
BOMBS	**BOMBING**	**VICTORIES**
100lb GP	Visual	0-0-0
FUSES		**FLAK**
.1 Nose, .025 Tail		None

198 Graz, AUT - Marshalling Yard 21/3/1945

GRAZ/KOFLACHER Marshalling Yard was attacked by PFF due to the cloud cover. Photos show hits South of the marshalling yard with direct hits on the South sidings probably cutting all through lines. A concentration also fell on the gas works and power station South of the yard. Only very slight damage to the KOFLACHER Marshalling Yard.

ESCORT	WEATHER	LOSSES
82FG	9/10	0-0-4
BOMBS	BOMBING	VICTORIES
500lb RDX	Visual	0-0-0
FUSES		FLAK
.1 Nose, .025 Tail		M-I-H

199 Vienna, AUT - Florisdorf Oil Refinery and Marshalling Yard 22/3/1945

This alternate target was bombed by PFF when the cloud covered primary could not be picked up by radar. A concentration of bombs fell on the M/Y and RR car works, scoring 16 strikes on buildings and car works and six strikes on marshalling yards. Four bombs fell on the South choke point probably cutting the through lines.

ESCORT	WEATHER	LOSSES
1FG	5/10	0-1-10
BOMBS	BOMBING	VICTORIES
500lb RDX	Pathfinder	0-0-0
FUSES		FLAK
.1 Nose, .025 Tail		I-A-H

200 Vienna, AUT - Kagran Oil Refinery 23/3/1945

Bombs fell West of the refinery. One bomb scored a direct hit on a storage tank just South of the refinery causing a huge explosion and fire. Four hits were scored on RR tracks West of the refinery, probably cutting the through lines.

ESCORT	WEATHER	LOSSES
1FG	CAVU	1-2-11
BOMBS	BOMBING	VICTORIES
500lb RDX	Visual	0-0-0
FUSES		FLAK
.1 Nose, .025 Tail		I-A-H

201 Ceske Budejovice, CZE - Marshalling Yard 24/3/1945

Bomb strike photos show a concentrated pattern of strikes across the entire yard scoring numerous hits on the loco depot, passenger station, goods wagons, and tracks, probably cutting the through lines.

ESCORT	WEATHER	LOSSES
None	CAVU	0-0-0
BOMBS	BOMBING	VICTORIES
100lb Clusters	Visual	0-0-0
FUSES		FLAK
.1 Nose, .025 Tail		None

202 Wels, AUT - Airdrome 25/3/1945	ESCORT 332FG	WEATHER CAVU	LOSSES 0-0-0
The primary target, PRAGUE Kbely Airdrome, had to be abandoned when the interphone between the pilot and bombardier went out. The WELS Airdrome was hit visually. A good concentration of bombs fell across the A/D continuing into the hanger area South of the A/D destroying 5 A/C and damaging three. Numerous hits to hangers and service aprons causing one explosion.	BOMBS 100lb Frags	BOMBING Visual	VICTORIES 0-0-0
	FUSES .1 Nose, .025 Tail		FLAK None

203 Straszhof, AUT - Marshalling Yard 26/3/1945	ESCORT 1FG	WEATHER CAVU	LOSSES 3-0-2
Concentration of hits across the West half of the M/Y and West choke points damaging goods wagons and cutting through tracks. Large fires were left burning among the goods wagons in the center of the marshalling yard.	BOMBS 100lb Frags	BOMBING Visual	VICTORIES 0-0-0
	FUSES .1 Nose, .025 Tail		FLAK S-I-H

204 Graz, AUT - Marshalling Yard 26/3/1945	ESCORT None	WEATHER 4/10	LOSSES 0-1-2
This was a four plane unescorted Lone Wolf mission. Bomb strike photos show a pattern of bombs across the South end of the station yard with hits on rolling stock and trackage. Hits were also scored on the Goods Yard North of the wagon repair facility, and across the South choke point in the main marshalling yard with hits on rolling stock and trackage. Three of the four planes were hit by flak, but all returned.	BOMBS 500lb RDX	BOMBING Visual	VICTORIES 0-0-0
	FUSES .1 Nose, .025 Tail		FLAK M-A-H

205 Linz, AUT - Benzol Plant 31/3/1945	ESCORT 325FG	WEATHER 9/10	LOSSES 1-0-4
This was a PFF mission, which bombed through an overcast that almost entirely obscured the target. Some hits were visible in the North end of the main marshalling yards. Six hits were visible near the choke point. One string of bombs walked across the yard.	BOMBS 1000lb GP	BOMBING Pathfinder	VICTORIES 0-0-0
	FUSES .1 Nose, .025 Tail		FLAK M-A-H

Six aircraft were lost during the month due to mid-air collisions, flak, weather, and mechanical problems. Twenty-seven men were KIA, nineteen wounded and 10 POWs.

April 1945

The group flew 18 combat missions in its final month of the war. Combat planes logged 4,113 combat hours. There were 624 planes over target of the 648 who took off. Of the 1,422 bombs airborne, 1,244 hit their targets. Bombing accuracy was much improved with an average score of 68.9 per cent. This score was good enough for the group to rank sixth in the Air Force, up from thirteenth the previous month.

206 St Polten, AUT - Marshalling Yard 2/4/1945	ESCORT 31FG	WEATHER CAVU	LOSSES 0-0-0
Bomb strike photos show a very concentrated pattern of bombs on the M/Y scoring hits on trackage and goods wagons. The West choke point received 7 direct hits and 15 hits fell on the station yards.	BOMBS 500lb RDX	BOMBING Visual	VICTORIES 0-0-0
	FUSES .1 Nose, .025 Tail		FLAK None

207 Brescia, ITA - Marshalling Yard 5/4/1945	ESCORT 52FG	WEATHER 3/10	LOSSES 0-0-0
Good pattern of bombs across target area and West half of the freight yards scoring numerous hits on goods wagons and trackage. A number of goods wagons were destroyed and all through lines were cut. One large explosion and three fires were seen among the goods wagons. West and East choke points received direct hits and are probably destroyed.	BOMBS 500lb RDX	BOMBING Visual	VICTORIES 0-0-0
	FUSES .1 Nose, .025 Tail		FLAK None

208 Brescia, ITA - Breda Small Arms Works 6/4/1945

Heavy concentration of bombs completely covering target area, scoring numerous direct hits and near misses on all installations. Three workshops and two previously damaged workshops destroyed. Roofs of 1 large and 1 small assembly building destroyed. Other miscellaneous buildings severely damaged.

ESCORT	WEATHER	LOSSES
31FG	CAVU	0-0-0
BOMBS	BOMBING	VICTORIES
500lb RDX	Visual	0-0-0
FUSES		FLAK
.1 Nose, .01 Tail		None

209 Gorizia, ITA - Marshalling Yard 8/4/1945

Being briefed to only attack the primary target visually and finding it cloud covered, the group attacked Gorizia with a concentration of bombs falling across the M/Y with numerous hits on the rolling stock and tracks. Workshops on West side were hit directly at least five times. Through lines were probably cut. There were direct hits on rolling stock and tracks, on warehouses and the North choke point.

ESCORT	WEATHER	LOSSES
52FG	2/10	0-0-0
BOMBS	BOMBING	VICTORIES
250lb GP	Visual	0-0-0
FUSES		FLAK
.1 Nose, .01 Tail		None

210 Lugo, ITA - (Blue Force) - Troop Concentrations 9/4/1945

As they approached the release point another formation flew under them, forcing the Blue Force had to refrain from dropping on the target. As the allowed time to complete bombing was too close to go around for another run, they were forced to return their bombs to the base.

ESCORT	WEATHER	LOSSES
None	CAVU	0-0-0
BOMBS	BOMBING	VICTORIES
100lb Frags	Did Not Bomb	0-0-0
FUSES		FLAK
100lb Frag Clusters		H-S-I

210 Lugo, ITA - (Red Force) - Troop Concentrations 9/4/1945

Everything went exactly as planned for the Red Force allowing them to bomb with a good concentration and excellent coverage of the target areas. Bomb smoke prevented assessment of the results.

ESCORT	WEATHER	LOSSES
None	CAVU	0-0-1
BOMBS	BOMBING	VICTORIES
100lb Frags	Visual	0-0-1
FUSES		FLAK
100lb Frag Clusters		S-I-H

211 Lugo, ITA - (Red Force/Blue Force) - Troop Concentrations 10/4/1945

Target attacked visually with good concentrations and excellent coverage of target areas.

ESCORT	WEATHER	LOSSES
None	CAVU	0-0-0
BOMBS	**BOMBING**	**VICTORIES**
100lb Frags	Visual	0-0-0
FUSES		**FLAK**
100lb Frag Clusters		S-I-H

212 Bronzolo, ITA - Marshalling Yards 11/4/1945

Photos show a concentrated pattern falling across the center of the M/Y and North and South choke points. Thirty strikes among goods wagons and trackage in yard. Seven direct hits on South choke point. Four direct hits on warehouses West of the yard. One direct hit or very near miss on RR bridge South of the yard.

ESCORT	WEATHER	LOSSES
1FG	CAVU	1-0-16
BOMBS	**BOMBING**	**VICTORIES**
500lb RDX	Visual	0-0-0
FUSES		**FLAK**
.1 Nose, No Delay Tail		M-A-H

213 St. Veit, AUT - Railroad Bridge 12/4/1945

Concentrated pattern of bombs fell across the bridge with four direct hits or very near misses. Both North and South approaches received direct hits with RR lines cut. RR line East of RR bridge received one direct hit or very near miss. East approach of highway bridge West of target received one direct hit.

ESCORT	WEATHER	LOSSES
332FG	5/10	0-0-0
BOMBS	**BOMBING**	**VICTORIES**
1000lb GP	Visual	0-0-0
FUSES		**FLAK**
.1 Nose, .025 Tail		None

214 Malcontenta, ITA - Ammunition Factory 14/4/1945

Only one box of four was able to identify the target but they scored 67.4% with hits on North and South filling plant, across the ammo storage center, scoring direct hits on both buildings of the filling plant and storage buildings South and North-east of the target area. Dock facilities received several hits and small explosions occurred.

ESCORT	WEATHER	LOSSES
1FG	9/10	0-0-0
BOMBS	**BOMBING**	**VICTORIES**
500lb RDX	Visual	0-0-0
FUSES		**FLAK**
.1 Nose, No Delay Tail		H-S-I

215 Bologna, IT - Troop Concentrations (Red) 15/4/1945	ESCORT	WEATHER	LOSSES
Excellent coverage of the entire area with several explosions visible. Many buildings were destroyed by direct hits and a considerable number of ground defenses and slit trenches also received direct hits. Many hits were also observed on the highway and highway bridge.	None	4/10	0-0-1
	BOMBS	BOMBING	VICTORIES
	500lb RDX	Visual	0-0-0
	FUSES		FLAK
	.1 Nose, .025 Tail		S-I-H

215 Nervessa, ITA - Railroad Bridge Diversion (Blue) 15/4/1945	ESCORT	WEATHER	LOSSES
Photos show a heavy concentration of bombs covering the railroad diversion and cutting the tracks in numerous places. Three direct hits were probably scored on the bridge. Heavy concentrations also covered the road diversion and the rail lines South of the bridge.	52FG	8/10	1-0-0
	BOMBS	BOMBING	VICTORIES
	500lb RDX	Visual	0-0-0
	FUSES		FLAK
	.1 Nose, .025 Tail		None

216 Bologna, ITA - Troop Concentrations 16/4/1945	ESCORT	WEATHER	LOSSES
The group was assigned to bomb enemy troop concentrations in the BOLOGNA area, but cumulus clouds to 10,000 feet obscured the area, forcing them to return their bombs to the base.	332FG	CAVU	0-0-0
	BOMBS	BOMBING	VICTORIES
	500lb RDX	Did Not Bomb	0-0-0
	FUSES		FLAK
	.1 Nose, .025 Tail		S-I-H

217 Avisio, ITA - Viaduct and Railroad Diversion 19/4/1945	ESCORT	WEATHER	LOSSES
The target was attacked by five bomb groups including the 461st. Bomb strike photos show a concentration of hits across the main bridge and diversion though an effective smoke screen prevented any further observation of the results.	1FG	HAZE	0-0-0
	BOMBS	BOMBING	VICTORIES
	500lb RDX	Visual	0-0-0
	FUSES		FLAK
	.1 Nose, No Delay Tail		M-I-H

218 Lusia, ITA - Bridge 20/4/1945

Target was attacked visually. Photos show a heavy concentration smothered the target. Many direct hits were scored on the bridge and its abutments and one southern span of the bridge was completely destroyed. The highways at both ends were heavily cratered and the building area at the South end of the bridge is completely destroyed.

ESCORT	WEATHER	LOSSES
332FG	CAVU	0-0-0
BOMBS	**BOMBING**	**VICTORIES**
1000lb RDX	Visual	0-0-0
FUSES		**FLAK**
.1 Nose, .025 Tail		S-I-H

219 Attnang, AUT - Puchiem Marshalling Yard 21/4/1945

Excellent concentration across center of yard with direct hits on rolling stock and trackage. South choke point and Northwest area well hit. Roundhouse and large warehouse at Northwest side of yard received several direct hits causing four separate explosions. Highway overpass at South end of yard cut in two places.

ESCORT	WEATHER	LOSSES
332FG	3/10	0-0-1
BOMBS	**BOMBING**	**VICTORIES**
500lb RDX	Visual	0-0-0
FUSES		**FLAK**
.1 Nose, .025 Tail		None

220 Badia, ITA - Road Bridge 23/4/1945

South two thirds of the bridge was demolished and laying in the river. Both highways on either side of the river were cratered.

ESCORT	WEATHER	LOSSES
14FG	CAVU	0-0-1
BOMBS	**BOMBING**	**VICTORIES**
1000lb RDX	Visual	0-0-0
FUSES		**FLAK**
.1 Nose, .01 Tail		M-A-H

221 Rovereto, ITA - Marshalling Yard and Gun Emplacements 24/4/1945

Heavy concentration of hits across the yard with 45 strikes scored on goods wagons and trackage. The North and South choke points and through tracks probably cut by direct hits.

ESCORT	WEATHER	LOSSES
None	CAVU	0-0-1
BOMBS	**BOMBING**	**VICTORIES**
500lb RDX	Visual	0-0-0
FUSES		**FLAK**
.1 Nose, .025 Tail		S-I-H

222 Linz, AUT - Marshalling Yard 25/4/1945	ESCORT 305FW	WEATHER CAVU	LOSSES 2-1-10
Group attacked target by visual method. Concentrated pattern across the target area, but target obscured by smoke from previous bombing.	BOMBS 500lb RDX	BOMBING Visual	VICTORIES 0-0-0
	FUSES .1 Nose, .025 Tail		FLAK I-A-H

223 Linz, AUT - Marshalling Yard 26/4/1945	ESCORT 305FW	WEATHER CAVU	LOSSES 0-0-0
Target of opportunity, bombed visually. Concentrated pattern of bombs fell on Northeast portion of marshalling yard with a number of hits on tracks and rail installations. Two large warehouses were severely damaged by direct hits. 2nd concentration fell on Southeast portion of M/Y damaging a number of goods wagons. All through lines probably cut.	BOMBS 250lb Clusters	BOMBING Visual	VICTORIES 0-0-0
	FUSES .1 Nose, .025 Tail		FLAK None

The group only lost two ships to combat, both to flak. German fighter did not challenge the group on any of their missions. In the final month of the war the group suffered 8 men killed, one of which was 2nd Lt. William Jones, the Co-pilot aboard Miss Lace who was murdered after successfully bailing out of his crippled ship over Austria on 25 April, 1945. Seven men were wounded and 6 became POWs. Twenty new crews joined the group only to see the end of the war.

THE HARD MISSIONS

ROMANO AMERICANA OIL REFINERY:
Ploesti, Romania - July 22, 1944

THE FIFTEENTH AIR FORCE attacked the oil refineries at Ploesti Romania for the second time in a week, sending 163 B-17s and 490 B-24s against the targets. The 461st Bomb Group, consisting of 26 aircraft, was assigned the Romana Americana Oil Refinery. Twenty two aircraft bombed the refinery by PFF with 1000 lb. bombs. The lead aircraft was hit by flak and bombed short. Other ships who bombed on the lead also bombed short. Other ships, who waited and bombed with the Deputy Lead, hit the target successfully.

One other aircraft attacked the Caprija Bridge, while three others were early returns as a result of gas leaks. Two ships were lost to flak, while two others bailed out their crews near the base. Seventeen other planes were holed by flak.

AAF#	Type	Group	Sq.	Sq. #	Nickname
41-29362	B-24H	461st	764th	7	Lucky Seven

PILOT: 2nd Lt. Clarence W. Bloxom, San Antonio, TX, Married
Missions: 2
LAST SIGHTED: 11:25, Between the IP and target
WHERE: Crash landed 3 miles east of Alexandria, Romania

Prior to the IP Lt. Bloxom was having trouble keeping up with the formation. He made a slow turn to the right and dropped his bombs. The ship was last seen with one engine feathered. Lt. Bloxom later successfully crash landed his plane 3 miles west of Alexandria, Romania. The pilot, 2Lt. Bloxom; engineer, S/Sgt. James M. Jordan; bombardier, 2Lt. William R. Braun; nose gunner, Cpl. Albert I. Paul; and ball gunner, Cpl. Richard R. Bryant were still aboard when it crash landed. The rest of the crew had previously bailed out. The navigator, 2Lt Carter M. Armstrong, Jr. was killed when he bailed out of the aircraft, and apparently slipped out of his parachute harness.

AAF#	**Type**	**Group**	**Sq.**	**Sq. #**	**Nickname**
42-50948	B-24J	461st	766th	Unk	None

PILOT: 2nd Lt. Elias E. Moses, Arnold, PA, Single

Missions: 2

LAST SIGHTED: 11:30, over the target

WHERE: Disappeared during bomb run over Ploesti, Romania

The formation spread out to avoid intense flak following the bomb release. It was during this time the aircraft flown by Lt. Moses disappeared. Little is known about the circumstances of the lost ship, although it was noted Lt. Moses was out of formation 10 minutes after the attack. The crew was eventually returned to duty.

AAF#	**Type**	**Group**	**Sq.**	**Sq. #**	**Nickname**
41-29313	B-24H	461st	766th	45	Hottest ??? in Town

PILOT: 2nd Lt. Turner M. Holmes

Missions: Unk

LAST SIGHTED: Bailed out his crew over the base

WHERE: Torretta Field, Italy

One of 17 aircraft hard hit by flak over the target, Lt. Holmes was able to nurse his ship back to Torretta Field when leaks in his gas line forced him to bail out his crew near the field.

AAF#	Type	Group	Sq.	Sq. #	Nickname
42-78253	B-24G	461st	765th	29	None

PILOT: 2nd Lt. Noble Taylor
Missions: Unk
LAST SIGHTED: Bailed out his crew over the base
WHERE: Torretta Field

Also hard hit by flak over the target, Lt. Taylor was able to return his crew to the base. But they were compelled to bail out near the base with only one engine still running. F/O Irving Smithkin, navigator, was fatally injured in parachuting to earth.

HERMAN GOERING TANK WORKS:
Linz, Austria - July 25, 1944

THE HERMAN GOERING TANK WORKS was a huge combination of coking plants. Blast furnaces, steel works, and chemical works, complete with power station, rolling mill, stamping and press shops. The Fifteenth Air Force would send 399 heavy bombers and 198 fighters to drop 1111 tons of bombs on the site.

German fighter defenses were estimated to be 25/30 single engine fighters that would oppose the formation in the Adriatic area, either enroute to or from the target. They could also anticipate 75 single engine fighters and 60/75 twin engine aircraft in the target area.

Flak batteries were expected to be 164 heavy guns in the Linz area, 24 guns in Wels, 22 heavy guns in Enns, and another 70 guns in Steyr. It was expected that smoke screen would also be deployed.

The Fifteenth would deploy six B-17 groups from the 5th Wing, three B-24 groups from the 49th Bomb Wing, four B-24 groups from the 55th Wing, and four more B-24 groups from the 304th bomb wing, with the 49th Bomb Wing leading. Each aircraft would carry 1,000lb GP bombs with .1 second delay nose fuses and .025 second delay tail fuses. Because they would be bombing by PFF each group would be led by Pathfinder aircraft.

The 49th Wing would fly a staggered column of groups stepped up; 484th Bomb Group first, 461st Bomb Group second, and 451st Bomb Group third. Time over the target would be 1100. The 484th would bomb from 22,000 feet, 461st from 23,000 feet, and the 451st from 24,000 feet. Rally after bombs away would be left to Kresmunster with a direct route back to base.

Due to the length of the mission, it was determined the 461st would not lower their ball turrets until after the IP, reducing drag and conserving fuel. However, the decision would have disastrous consequences. The reaction of the German fighters was heavy and aggressive. The 461st was attacked by approximately 150 aircraft between the IP and target and was executed in four phases. "The first attack was made by FW 190s and Me 109s attacking in units of 20 aircraft abreast. The second these flights were broken into smaller units. The third attack was made by individual single engine fighters. The fourth attack was made by rocket carrying Me 410s. The attacks were generally made from 3 to 9 o'clock, front to rear, except for the Me 410s which attacked from the rear."

The German battle plan called for organizing their fighters into two strong forces. The larger force was held in the Linz area, while the other was sent along the bombers incoming route. The smaller German force engaged the 49th Bomb Wing escorts leaving the bombers with little or no protection. At that point, the larger German force was able to attack the unescorted Liberaiders knocking down eleven aircraft over the target. In addition to those lost over the target, Lt. Douglas Herrin, one of eight that was able to return to Torretta, was so badly battle damaged that his ship was salvaged. Lt. Casper Jenkins, with three wounded aboard, washed out attempting to land at Foggia. Lt. Edgar Trenner, using parachutes in place of flaps and landing with a punctured tire, washed out at the base. Lt. Robert Wester was forced to bail out his crew over the Isle of Vis. Of the 113 officers and enlisted men who went down on the mission, 7 officers and 9 enlisted men were on their 50th sortie. Lt. Ernest Henry was the only man on his 50th sortie to return to the base. American gunners claimed 31 German fighters shot down, 19 more probably shot down and 9 damaged. It was considered likely that many of the probables had been shot down, but the gunners were too busy to follow the crippled fighters to the ground.

The 461st put up 26 aircraft of which 5 were early returns and 15 were either lost over the target or attempting to land.

AAF#	Type	Group	Sq.	Sq. #	Nickname
41-28850	B-24H	461	766	44	None

PILOT: 1st Lt. Joseph B. Hesser, North Vernon, Indiana, Single, KIA
Last Sighted: Linz, Austria
Where Crashed: Crashed into a farmhouse near the village of Luftenberg outside of Linz causing it to burn down.

#44 was hit in the rear bomb bay, causing a large fire which gushed from the left rear bomb bay in high volume. Two men bailed out of the right front bomb bay, one out of the waist window, and one from the camera hatch. After that it peeled to the left out of the formation seemingly still under control. It then slipped over and went into a spin. At some point two more men were able to bail out. #44 then fell approximately 5,000 feet below the formation before exploding, the fuselage breaking in two and the wings falling off. The pilot, 1st. Lt. Joseph B. Hesser, right waist gunner S/Sgt. Robert H. Kimble, top turret gunner S/Sgt. Coy Westfall, Jr. and tail turret gunner Cpl. George T. Tyler were killed in the explosion. The six survivors became POWs.

AAF#	Type	Group	Sq.	Sq. #	Nickname
42-52451	B-24H	461st	767th	66	None

Pilot: 2nd Lt. John J. Kane, Chicago, Illinois, Married
Last Sighted: Between Unt. Wieszenbach and Linz, Austria
Where Crashed: Breitenau near Molln, greater Buchberg

#66 was hit by flak either in the bomb bay or near the inboard engine. The plane was burning and fell out of the formation, down and to the right. It appeared to be still under control. S/Sgt. Frank G. Rogan, the right waist gunner, had his right leg blown off by 20mm cannon fire as well as flak wounds and internal injuries. In spite of this he was still able to bail out. He later died in Lt. Kane's arms enroute to a hospital after receiving no medical help and slow transportation to the hospital. S/Sgt. Harold F. Williamson, tail turret gunner, was killed when 20mm cannon fire blew the turret apart. S/Sgt. Earl R. Roberts, ball turret gunner, was able to bail out, but was strafed by a German fighter while still going down in his chute. Lt. Koistinen later found him hanging from a tree still in his chute. He back was full of machine gun holes.

AAF #	Type	Group	Sq.	Sq. #	Nickname
42-95257	B-24H	461st	767th	72	None

Pilot: 2nd Lt. Rolland T. Olson, Iron Mountain, Michigan, Married
Last Sighted: 1105 between Unt. Wieszenbach and Linz, Austria
Where Crashed: Farrach, District Judenburg/Steiermark

After a second pass by fighters, #72 had flames coming from the #3 and #4 engines forcing the crew to bail out. The aircraft passed under the formation, slid over on its right wing, and exploded. The entire crew was captured by German troops and became POWs. They were later liberated and returned to duty.

AAF#	Type	Group	Sq.	Sq. #	Nickname
41-28836	B-24H	461st	767th	61	None

Pilot: 2nd Lt. Robert H. Fisher, Idaho Falls, Idaho, Single
Last Sighted: 1105 between Unt. Wieszenbach and Linz, Austria
Where Crashed: A few miles south of Linz, Austria

Shot down by fighters. The nose of the aircraft was riddled by 20mm cannon fire during head on attacks. 2nd Lt. Edward M. Mancourt, bombardier, William Jelens, Jr., top turret gunner, and S/Sgt. Anthony A. Lucero, the nose gunner were killed in the attack. However, Lucero was also credited with shooting down one of the attacking German aircraft. The other eight men, including a photographer, bailed out of the blazing bomber and became POWs.

AAF #	Type	Group	Sq.	Sq. #	Nickname
42-94732	B-24H	460th	767th	68	Tail Dragon

Pilot: Erwin W. Bowyer, Oklahoma City, Oklahoma, Married
Last Sighted: 1105 between Unt. Wieszenbach and Linz, Austria
Where Crashed: Between Unt. Wieszenbach and Linz, Austria

When the fighters attacked the formation one fighter raked Tail Dragon from nose to tail setting it afire. Fire was coming from the open bomb bays and streaming from the waist windows. The ship rolled over and flew on its back, falling through the formation and breaking it up. A few seconds later it blew up. Five of the crew were able to bail out and were eventually evacuated to the United States. The other five rode what was left of the plane to the ground and were declared officially dead.

AAF#	**Type**	**Group**	**Sq.**	**Sq. #**	**Nickname**
41-28693	B-24H	461st	764th	1	War Eagle

Pilot: 1st Lt. Richard E. Freeman, Butler, Pennsylvania, Single
Last Sighted: 48o 17' N 14o 19' E
Where Crashed: On Zauchersteg near St Peter in Au (8 km east of Steyer, Austria)

Four FW-190s made a pass at the ship starting a fire in the forward bomb bay and forcing the plane to drop out of formation. Aboard the aircraft T/Sgt. James F. Hagen, the right waist gunner was laying at his position dead from machine gun and 20mm canon fire. T/Sgt. Yancey L. Fulcher, Jr., the top turret gunner, also died at his position firing until he was killed by machine gun fire through the neck. The remaining eight members of the crew were able to bail out successfully although 1st Lt. Armanta R. D'Amante, bombardier, was wounded in his left leg; S/Sgt Willie Gibson, top turret gunner, was wounded in the back and legs; and S/Sgt. Jose M. Salas, nose turret gunner, was wounded in the head, back and legs. All became POWs.

AAF#	**Type**	**Group**	**Sq.**	**Sq. #**	**Nickname**
42-95383	B-24H	461st	764th	12	Lazy Lady

Pilot: 1st Lt. Grover F. Mitchell, Roanoke, Texas, Single
Last Sighted: 48o 17' N 14o 19' E
Where Crashed: Mistelbach Notmuehle Community, Upper Danube

Attacked by flights of Me-109s and FW-190s, Lazy Lady was one of three ships from the same box to be shot down simultaneously when one engine was shot out, forcing it out of the formation. The ship dove 22,000 feet until crashing to the earth in flames. Sgt. John L. Kennedy, nose gunner, was killed engaging the enemy when hit by 20mm cannon fire, which also blew off the nose of the plane. Both Lt. Grover F. Mitchell (pilot) and Lt. Herbert H. Hartford (co-pilot) were killed on the flight deck during a frontal attack by German fighters. Lt. Hartford had only been married for one month. S/Sgt. Ira W. Hebert was in his position as the top turret gunner when he was killed by cannon fire. S/Sgt. Lynwood C. Harrell, right waist gunner was observed lying in the waist by his gun. He had been hit by 7.9 machine gun and 20mm cannon fire. Replacing his oxygen mask had no effect. All efforts to help Sgt. Harrell failed. S/Sgt. Donald D. Robinson, right waist gunner and radio operator, was last observed seemingly unhurt, standing in front of the escape hatch with his chute on. For unknown reasons he failed to jump and rode the plane to the ground with his dead crewmates. This was his first mission. Three men, 1st Lt. Walter J. Roeben, navigator, Sgt. Donald G. Charles, ball turret gunner, and Jasper D. Trout, tail gunner, were able to bail out. They became POWs.

AAF#	**Type**	**Group**	**Sq.**	**Sq. #**	**Nickname**
42-52459	B-24H	461st	764th	8	Winona Belle

Pilot: 2nd Lt. Wray M. Stitch, Plainview. Minnesota, Single
Last Sighted: 48o 17' N 14o 19' E
Where Crashed: 48o 17' N 14o 19' E

The Winona Belle was attacked by FW-190s, causing a fire in the #2 and #3 engines, and forcing it to leave the formation in a spiral. The ship exploded prior to reaching the ground. Left waist gunner Sgt. Joseph R. Dalbec was discovered dead laying below his gun covered with blood. Sgt. Gene C. Dahike, right waist gunner, was found lying on the camera hatch. There had been several 20mm cannon hits in the waist section and he looked to be beyond help. Ball turret gunner Cpl. Julian R. Ducray was at his station at the time of the attack, but his guns were never heard by the rest of the crew. When the tail gunner came forward to bail out he reported the ball turret could not be seen due to the smoke and flames in the ship. It was presumed he never left the turret. 1st Lt.

James E. Harris, co-pilot, remained at the controls while the crew was bailing out. As the pilot's controls were inoperative Lt. Harris told him to leave the ship. It was unknown whether Lt. Harris left the ship or not. He was not seen by the crew once on the ground. F/O William M. Albright, bombardier, was believed to have left the ship. The nose gunner stated that F/O Albright was not in the nose when he left, and that the nose wheel doors were open so he assumed Albright had left that way. There was a lot of shooting in the area where the crew landed so it was believed F/O Albright may have been killed by German troops. At least one other crewman reported being shot at before he was captured. The other five crew members successfully bailed out and were captured by German troops. They were later evacuated back to the Unites States.

AAF#	Type	Group	Sq.	Sq. #	Nickname
41-28867	B-24H	461st	765th	21	Sleepy Time Gal

Pilot: 2nd Lt. Kenneth O. Githens, Portland, Oregon, Married
Last Sighted: 11:20, Linz, Austria
Where Crashed: 48o 17' N 14o 19' E

Sleepy Time Gal was attacked by five Me-109s in a pursuit curve. The left waist gunner on Lt. Arbuthnot's crew on ship #24 shot down the first Me-109. The other four fighters continued the attack on Sleepy Time Gal, setting fire to the bomb bay. It fell up in a slid and began losing altitude. The entire crew was able to bail out before the ship exploded. All the crew members were later liberated and returned home.

AAF#	Type	Group	Sq.	Sq. #	Nickname
42-52368	B-24H	461st	765th	34	Piecemaker

Pilot: 1st Lt. Glennial Fulks, Manfordville, Kentucky, Single, KIA
Last Sighted: 11:20, Linz, Austria
Where Crashed: Stranzberg, 22 km east of Linz, Austria

Fighters attacked Piecemaker setting it on fire. The wing was a solid sheet of fire with flames streaming back past the tail. The ship turned over on its back completely out of control. 2nd Lt. Warren O. Ludlum, the co-pilot, was able to bail out of the stricken plane. S/Sgt. George Diamond, right waist gunner, and S/Sgt. Peter Godino, tail turret gunner, were blown through a hole in the fuselage that had been caused by an explosion and survived. The three men became POWs and were later evacuated to the United States.

AAF#	**Type**	**Group**	**Sq.**	**Sq. #**	**Nickname**
41-29529	B-24H	461st	765th	23	None

Pilot: Robert A. Warren Jr., Weston, Mass., Single
Last Sighted: 11:20, Linz, Austria
Where Crashed: Hoersching, near Linz, Austria

#23 was attacked by German fighters and burst into flames. Fire was coming from the bomb bays and wing tanks. The ship pulled out of formation and salvoed the bomb load, after which it rolled over on its back and started down. Nine of the crew were able to abandon the plane. However, Cpl. Glen A. Myers, tail gunner, had been shot through the eye and was missing half his face. While he was able to get out of his turret, it was believed that he never bailed out and road the ship into the ground. The pilot, 2nd Lt. Robert A. Warren Jr., navigator 2nd Lt. Glenn H. Engbrook, and nose gunner Cpl. Lester J, Wilson, had been injured and rest of the crew were slightly burned, but otherwise okay.

* * * * *

In addition to the eleven ships lost over the target:

- 2nd Lt. Casper T. Jenkins (765th), flying #37, 42-52371, with three wounded men aboard, washed out his plane when attempting to land at Foggia. (It would be rebuilt and would resume combat operations, returning to the United States in June 1945).
- 1st Lt. Edgar M. Trenner (764th), flying #18, 41-28732, Swee' Pea, used parachutes in place of flaps, and landing with a punctured tire, washed out his plane at the base.

- 2nd Lt. Robert G. Wester (765th), flying #26, 42-52486, bailed out his crew over the friendly Isle of Vis in the Adriatic.
- 2nd Lt. Douglas A. Herrin (764th), flying #13, 44-40632, one of eight aircraft that returned to base, was so badly shot up it was salvaged. History would show there
- was worse in store for Lt. Herrin and his crew.

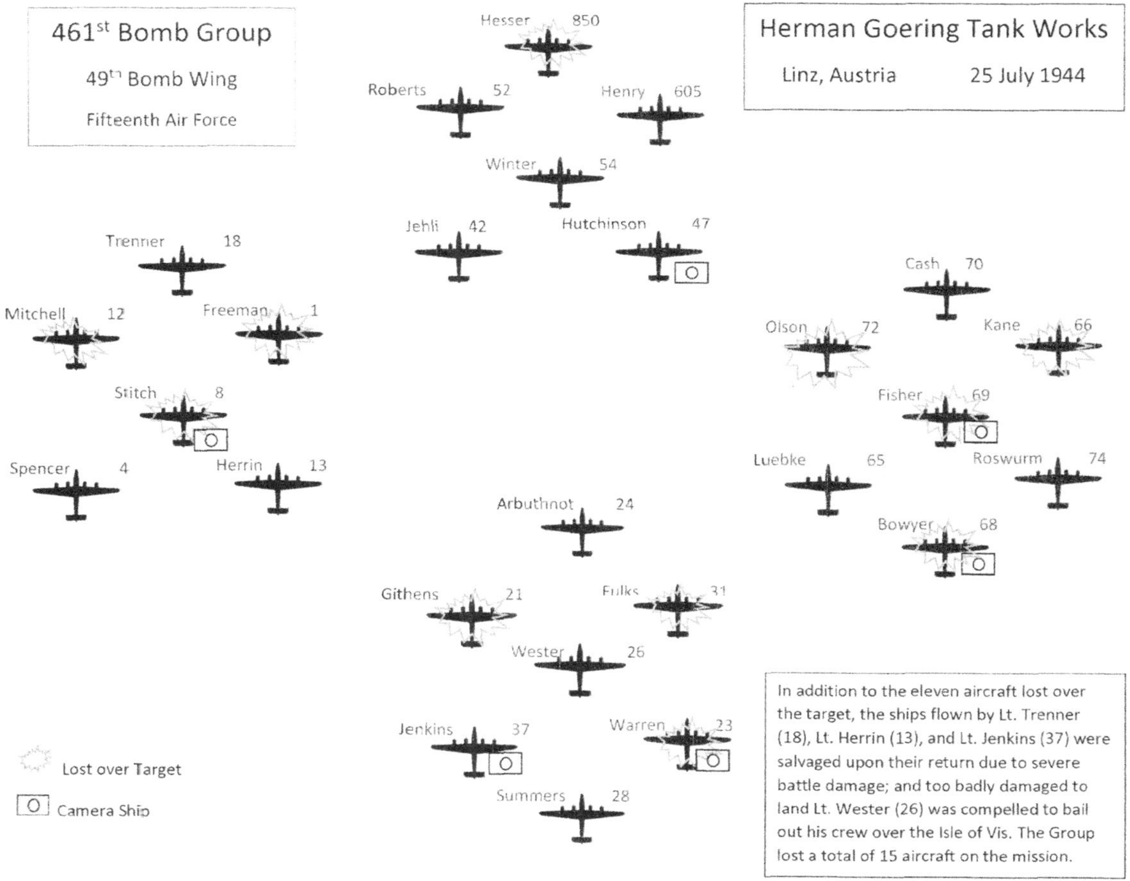
461st Bomb Group
49th Bomb Wing
Fifteenth Air Force
Herman Goering Tank Works
Linz, Austria
25 July 1944
Hesser 850
Roberts 52
Henry 605
Winter 54
Jehli 42
Hutchinson 47
Trenner 18
Mitchell 12
Freeman 1
Stitch 8
Spencer 4
Herrin 13
Cash 70
Olson 72
Kane 66
Fisher 69
Luebke 65
Roswurm 74
Bowyer 68
Arbuthnot 24
Githens 21
Fulks 31
Wester 26
Jenkins 37
Warren 23
Summers 28
Lost over Target
Camera Ship
In addition to the eleven aircraft lost over the target, the ships flown by Lt. Trenner (18), Lt. Herrin (13), and Lt. Jenkins (37) were salvaged upon their return due to severe battle damage; and too badly damaged to land Lt. Wester (26) was compelled to bail out his crew over the Isle of Vis. The Group lost a total of 15 aircraft on the mission.

MUNICH WEST MARSHALLING YARDS:
Munich, Germany - October 4, 1944

THE MUNICH MARSHALLING YARDS were a target of high strategic value. They were five miles long, running in an East-West direction. They included a station goods siding and repair shops in the western end of the yards, and sidings and junctions in the east. The maximum width was 490 yards. The attack would destroy a large concentration of military rolling stock, repair facilities, and would cut important rail lines passing through this important center.

It was expected the German Air Force could put up 125 single-engine fighters. Jet fighters had been seen in the area although no opposition was expected from Me-163's or Me-262s. The 305th Fighter Group provided general target cover, while the 14th Fighter Group escorted the 461st on penetration, target, and withdrawal. There were also 265 heavy flak guns defending the target.

The Fifteenth Air Force would put up 331 bombers, which dropped 791 tons of bombs on the target. Most groups bombed visually. The 49th Bomb Wing was first to attack the target, with the 451st in the lead at 23,000 feet and the 461st 6000 feet behind at 24,000 feet. The 461st had put up 28 aircraft. Two were early returns, leaving 26 to go over the target. Both groups deployed 26 aircraft in four boxes. They carried 500lb RDX bombs fused with .1 second delay nose fuses and .01 second delay tail fuses. The aircraft also carried 4 cartons of chaff, while would be dropped beginning two minutes from the IP and continuing throughout the attack. Following the 49th Bomb Wing was the 55th, 304th, and 5th Bomb Wings. Target time was 1200A.

Damage to the marshalling yards was extensive with concentrations at the East Entrance, center, and West Entrance. Many tracks in the central area were cut, warehouses and sheds were damaged, rolling stock was destroyed or damaged throughout the yards, the station siding was damaged, there were 15 craters in the station area, and 30 bombs hit the goods depot. The 461st, led by Lt. Col. Hawes, placed 70% of its bombs within 1,000 yards of the aiming point.

The flak was described as intense, very accurate, and heavy. Seven 461st ships were lost to flak, making it the worst loss of aircraft due to flak of any mission flown by the 461st. Sixteen to the remaining nineteen ships were holed by flak. There was no fighter resistance.

AAF#	Type	Group	Sq.	Sq, #	Nickname
42-78446	B-24J	461st	766th	47	Urgent Virgin

Pilot: 2nd Lt. Walter J. Chester, Montgomery, Alabama, Married, KIA
Last Sighted: 1200, Munich, Germany
Where Crashed: Sleisheim, Germany

Just as the Urgent Virgin turned off target it was hit when another ship alongside it blew up. It immediately caught fire and exploded. The pilot, Lt. Chester, co-pilot 2nd Lt. Frank J. Hagan. Bombardier 2nd Lt. Forrest E. Langley, navigator 2nd Lt. Milton Barth, top turret gunner Sgt. William L. Guthrie, tail turret gunner Sgt. Henry J. Winkler, nose turret gunner Sgt. Albert M. Bair, and left waist gunner S/Sgt. Ore B. Fitch were killed in the explosion. Their bodies were later found in a grave 5 miles south of Munich with a cross marking the spot. Right waist gunner S/Sgt. James P. Walsh became a POW, while ball turret gunner Sgt. Emory C. Clippert was hospitalized with injuries from the explosion.

AAF#	Type	Group	Sq.	Sq, #	Nickname
42-78444	B-24G	461st	765th	24	All American

Pilot: 1st Lt. Robert T. Chalmers, Rutland, Vermont, Single
Last Sighted: 1245, 50 miles north of Trieste, Italy
Where Crashed: In the Adriatic close to the Isle of Vis

There is some question about the All American, flown by 1st Lt. Robert T. Chalmers, and whether it was hit by the heavy flak, or developed engine trouble. Whatever happened the ship had two engines smoking badly and it was losing power. Finally unable to keep up with the formation by the time they were 50 miles north of Trieste, Italy, they decided to try for Yugoslavia. Once safely over Yugoslavia and outside the German area of control, Lt. Chalmers bailed out his crew where they were picked up and returned to the base. There was some question about whether or not the navigator, 1st Lt. Frank M. Dick had landed in a swampy area and drowned. While he was safely recovered from this mission, he did drown after bailing out of another ship on Dec. 15, 1944.

AAF#	Type	Group	Sq.	Sq, #	Nickname
42-51338	B-24J	461st	766th	54	None

Pilot: 2nd Lt. Robert D. Falkner, Fairview Village, Ohio, Married
Last Sighted: 1300, Zolta, Italy
Where Crashed: Vicinity of Locca, Italy north of Goritzia

42-51338 was lost to fighter attack, although the exact circumstances are not known. All ten member of the crew successfully bailed out. Sgt. Robert E. Davis, the ball turret gunner and S/Sgt. Floyd A. Bacon, the tail turret gunner were captured by German forces and held in a POW camp for the remainder of the war. Lt. Falkner and the remaining seven members of the crew were rescued by Italian partisans and brought back through enemy lines to Allied control.

AAF#	Type	Group	Sq.	Sq, #	Nickname
42-50970	B-24J	461st	644th	13	Lucky Lady

Pilot: 1st Lt. John L. Turner, Jr., East Orange, New Kersey, Married
Last Sighted: 1154, 48o 09' N 11o 30' E
Where Crashed: Approximately 5km south of Moosburg, Germany

Lucky Lady caught fire after being hit by flak over Munich, which forced the pilot, 1st Lt. John L. Turner to order the crew to bail out. Everyone in the front of the aircraft was able to bail out either through the nose wheel doors or the bomb bay. However, tail gunner Harold J. Bauer, left waist gunner William C. Foster, and right waist gunner Donald B. Harrison were still standing in the waist when the plane exploded. Sgt. Bauer and T/Sgt. Harrison were thrown from the ship when it came apart, but S/Sgt. Foster was unable to escape from the wreckage before it crashed, killing him instantly. Beside the one fatality, two men had broken legs and all had various lacerations, bruises, and strains. The surviving crew became POWs and were eventually returned to duty.

AAF#	Type	Group	Sq.	Sq, #	Nickname
44-41039	B-24J	461st	764th	1	Male Box

Pilot: 2nd Lt. William M. Powell, St. Elyria, Ohio, Single
Last Sighted: 1245, 49o 02' N 12o 34' E
Where Crashed: HBB-CR at Mining.

Little is known about the circumstances surrounding the fate of Male Box. It is known to have been hit by flak and exploded. The co-pilot, 2nd Lt. Lawton M. Jones, Jr. was killed instantly when a fragment from an exploding bomb struck him in the head. The remainder of the crew were able to bail out and were captured by German forces. They were later liberated and returned to duty.

AAF#	Type	Group	Sq.	Sq, #	Nickname
42-78247	B-24G	461st	765th	39	None

Pilot: 1st Lt. William E. Waggoner, Little Rock, Arkansas, Single
Last Sighted: 1202, Munich, Germany
Where Crashed: Near Landsberg, Germany about 10 miles SW of Munich,

Aircraft #39 was flying the #2 position in "A" Flight when it received a direct hit by flak. It peeled away from the formation with #1 and #3 engines on fire and smoking badly. Lt. Waggoner was attempting to reach Switzerland, but his rate of decent was too great forcing an emergency landing in the Landsberg area. German troops removed co-pilot Lt. Norman C. Schlarp, who was badly wounded by the flak shell, from the ship and left him on the ground. The other crew members were rounded up and taken to a camp. Lt. Schlarp was later picked up by someone, but died a few days later in a German hospital.

AAF#	Type	Group	Sq.	Sq, #	Nickname
44-40896	B-24	461st	764th	4	Unknown

Pilot: 1st Lt. Douglas A. Herrin, Sumrall, Mississippi, Single, KIA
Last Sighted: Over the target
Where Crashed: Over the target.

After the group dropped its bombs the heavy flak caused them to make a sharp right turn. In the formation before them, one of the planes had a bomb hung up on a bomb shackle. That bombardier pulled the salvo switch and the bomb dropped out. As the 764th Squadron turned right, the bomb hit 44-40896, completely destroying the plane and entire crew. Aboard the plane flying as co-pilot was William H. Tallant, the 764th Bomb Squadron Operations Officer who had just been promoted to Major that day. The rest of the crew included:

2nd Lt. Peter Konapaka	Bombardier
2nd Lt. James D. Wagg	Navigator
S/Sgt. Louis J. Kain	Radio Operator
S/Sgt. John J. Juran	Engineer
Sgt. Harold J. McCrary	Gunner
Sgt. Morris D. Worman	Gunner
S/Sgt. Robert C. Haserlat	Gunner
Sgt. William F. Balen	Gunner

On July 25, 1944 Lt. Herrin and his crew escaped death when their ship was so shot up over Linz, Austria that it was salvaged after he crash landed back at Torretta Field. Sadly, their luck ran out.

ODERTAL:
December 17, 1944

THE FIFTEENTH AIR FORCE sent 231 Liberators from the 49th and 304th Bomb Groups to bomb the Odertal Oil Refinery. Flying as escorts would be 93 P-38s and 207 P-51s from the 305th and 306th Fighter Wings. Aircraft would be loaded with 500# RDX bombs fused with .1 nose and mixed .01 and .025 tail fuses. PFF aircraft would lead all units for navigation and bombing. All units loaded 2 cartons of "window" and dispensed starting three minutes before IP (Initial Point) at the rate of three bundles every 20 seconds until clear of flak. The 451st and 461st will line rendezvous behind the 484th. Bombing altitude will be 24,000 feet for the 484th, 25,000 feet for the 451st, and 26,000 feet for the 461st.

Because of the German offensive in the Arden Forest and the Russian advances in the East, mission planners did not expect a great deal of trouble from German fighters. In fact, only night fighters were known to be present in the area. The Intelligence Annex for the mission stated "Estimate little or no opposition VIENNA line to BLECHHAMMER with possible 25-30 s/e ftrs tgt area believed drawn from training units and not first line opposition capacity as likely to be encountered in central GERMANY, and not the same fighting capacity as units which opposed our formations during the past summer"

However, the Luftwaffe understood the importance of protecting the oil refinery in spite of the heavy losses they had suffered in September. They had placed approximately 100 fighters in the area consisting of Bf-109-G10s and G14s, and the more heavily armored FW-190-7s, A-8s, and A-9s, which would be flown by well trained and motivated pilots. These pilots, using the cloud cover and accurate directions from ground controllers, would prove to be even more effective due to the late arrival of the American fighter escort.

Their first target would be the 49th Bomb Wing. The 461st had been briefed to not deploy their ball turrets until the IP to reduce the drag and conserve fuel for the long mission. As a result the ball turrets, which usually would account for the most aerial victories due to their computing sights, fired the least amount of ammunition and scored the least number of kills. The German aircraft, heavily armored FW-190s, sliced through the bomber formations, which soon began to fall apart. The fighters, slate gray and white with black crosses on the wings and fuselage, some with light stripes around the fuselage, were identified as belonging to JG-300, Gruppe II known as "Wilde Sau" (Wild Boar). A total of eighteen 24s, including 10 from the 461st, would be lost during the battle. The attacks against the bombers ended when American fighters, which had been delayed by bad weather, arrived to engage the enemy. The American fighters claimed 23 sure kills and another 9 probables against American fighter losses of only 4 P-38s and 2 P-51s. The bombers claimed another 26 kills, which does not include any claims by gunners from the aircraft that had been lost.

As a result of their losses the group bombed with only 15 aircraft in one attack unit of three boxes. A six ship front was flown. The lead ship overflew the IP and the briefed axis was overflown by 12°. Bombing was done by pathfinder. The IP was picked up by the navigator who turned the formation into the briefed heading. The pathfinder operator had no difficulty killing the drift which was 1 ½ degrees. The target was followed in very close to the center of the scope. Pathfinder team coordination was particularly good.

* * * * *

AAF#	**Type**	**Group**	**Sq.**	**Sq. #**	**Nickname**
42-50953	B-24J	461st	765th	39	Flying Finger

PILOT: 2nd Lt. Phillip J. Crossman, Lima, Ohio, Single

Missions: 10 **LAST SIGHTED:** 11:45, Muglitz **DOWN:** 12:00

WHERE: 8 miles SW of Kutzendorf, CZ (CZ Poloce) 2km south Neutitso-neutitschsin Machrisch, Welka Weisengard, Kutzendorf, CZ. Mahr. 4 bodies in wreckage.

The "Flying Finger" was flying number three position. (Capt. Robert T. Chalmers was flying the number two position.) Capt. Jack Riley, gunnery officer, was in the upper turret when the B-24 left the formation. According to Capt. Chalmers the formation was hit by 50 plus fighters in the Muglitz area, Sgt. Bernard C. Freeman, tail gunner on the Flying Finger" said, "Got him!", but it sounded like he himself was badly hurt. Sgt. Freeman shot down an ME-109. The ball turret was up at the time of the attack and the crew was unable to get into action before the order to bail out came. Enemy fighters crisscrossed underneath the bomber almost at will. "Tail shot up bad, controls were very bad. Ship in very bad shape." The navigator said, "I gave him (Crossman) a heading to fly towards the Russian lines, then the interphone went out." The aircraft was riddled and went down fast. The crew started to bail out at 22,000, last one out at 1,500. Ball, tail, both waist gunners went in with the bomber. Most of the crew was rounded up by the Germans at 2PM, Lt. Crossman was not captured until 9:50PM 12/18/44.

AAF #	**Type**	**Group**	**Sq.**	**Sq. #**	**Nickname**
42-52025	B-24J	461st	765th	26	Arsenic and Lace

PILOT: 2nd Lt. Gerald Robert Smith, Grand Rapids, MI, Married
MISSIONS: 11
LAST SIGHTED: 11:45, Muglitz
DOWN: 12:00
WHERE: Co-pilot says 9 minutes from target. 12km south of Presenitz on NE tip of CZ.

This crew flew 11 missions between October 17 and December 17, 1944 including Vienna and Brux. #26 was hit in the number three engine by fighters. Number three was on fire, but the aircraft was still under control. The nose gunner, Cpl. David D. Brewer, nose gunner, comment on enemy rockets. Fighters crisscrossed and set bomb bays on fire. 2nd Lt. Arthur H. Carlson, bombardier, motioned for Smith to release the bombs. At this time Lt. Carlson was right below the upper turret. Sgt. Morris R. Goldman, upper gunner, called out, "Enemy fighter, six o'clock level!" The fighters crisscrossed from both sides of the plane. Left waist gunner Edwin L. Howard fired at the fighters, and was killed by a 20mm shell through the chest. The bomber's oxygen system was punctured. Lt. Carlson, who was 6'4", 180lbs, blue eyes, light brown hair, was trying to release Sgt. Goldman's jammed seat release without oxygen. Lt. Carlson freed Sgt. Goldman from the upper turret, but both collapsed from a lack of oxygen at the flight deck entrance. Navigator Lt. Milton A. Klarsfeld helped nose gunner David D. Brewer from his turret. The nose wheel doors would not open and the exit to the bomb bays was blocked by the unconscious bodies of Lt. Carlson and Sgt. Goldman. #26 fell off and to the right, and went under Capt. Chalmers aircraft. A wing broke off---fuel tanks exploded. 2nd Lt. Warren S. Steinhauser, 765th saw 10 men bail out of #26---he saw no chutes open. Capt. Robert T. Chalmers polled his crew and determined at least 10 men were seen going out of #26. The Germans found five bodies in the wreckage. Carlson, Goldman, Brewer, Abramson, and Howard. 2nd Lt. Vrooman L. Francisco, co-pilot met Phillip Crossman's crew at Stalag Luft One.

AAF #	Type	Group	Sq.	Sq. #	Nickname
44-41016	B-24J	461st	765th	35	Unknown

PILOT: 2nd Lt. Thomas K. West (KIA), Volin, South Dakota, Single
MISSIONS: Unknown
LAST SIGHTED: 11:45, Muglitz
DOWN: 12:30
WHERE: 49-27′ N 17-27′ E Prerov, CZ

2nd Lt. Donald J. Rhodes (765th), "Lt. West's plane pulled out of formation and were below us about 1,000 feet. At the time there were several enemy fighters around them, and while I watched I saw Lt. West's crew shoot down three of the enemy fighters. We passed them trying to catch the formation, and they were out of sight behind our right wing." Sgt. Trefry A. Ross, right waist on West's plane, "We were hit by enemy fighters about 12 o'clock, 5 minutes from the target. Our aircraft burst into flames." Sgt. Thomas E. Deibert, top turret gunner was killed by fighters. 2nd Lt. Thomas M. Noesges, bombardier, was struck in the head by a fragment from a German shell, and had no recollection of getting out of the doomed bomber. Lt. Thomas W. Qualman, navigator, "We were shot down about noon on our way north toward Germany." CZ women: "One plane fell in Troubkach in Prerova; there six are buried. The other fell in Lesne u Vel Mezirrici." West, Deibert, Doe, Yesia, Mergo and Gaul died in this action.

AAF #	Type	Group	Sq.	Sq. #	Nickname
42-95304	B-24J	461st	766th	42	Unknown

PILOT: 2nd Lt. Nicholas Sidovar, Algha, New Jersey, Single
MISSIONS: Unknown
LAST SIGHTED: Unknown
LAST RADIO CONTACT: 20 minutes after target time, just south of Bratislava, CZ at 14,300 feet
DOWN: 14:15 - 15:00, 20 miles SW of Agram Karloraw, Karlo? Oglj, Hungary
Weather: 8/10 cumulus at 16,000

Aircraft hit by fighters just before target. Tried to feather an engine just after leaving target and dropped out of formation. 20 minutes after target time 1st Lt. Jay M. Garner contacted Lt. Sidovar on "A" channel. Lt. Sidovar was trying to contact fighter escort. Sidovar told Lt. Garner that the engine would not stay feathered and was vibrating badly. There was also a slight gas leak. Lt. Sidovar finally contacted fighters and gave his position as just south of Bratislava, CZ at 14,300 feet. He was holding altitude. Entire crew returned to the states.

AAF #	Type	Group	Sq,	Sq. #	Nickname
42-51319	B-24J	461st	767th	69	Welcome Wagon

PILOT: 2nd Lt. Frederick B. Capalbo, Greenwich, CT, Single
DOWN: 12:00
MISSIONS: Perhaps first; left waist gunner, John A. Villars, first. Co-pilot William C, Rundbaken had 33 missions.
LAST SIGHTED: 12:00, Muglitz   Capt. Miller's flight
WHERE: Prevov, CZ Cockery, CZ Preavvie, CZ

#69 left the formation 10-15 minutes from IP and was 1,000 yards behind the formation when enemy fighters struck. S/Sgt. Robert F. Miller, tail gunner on lead ship, "#69 was attacked by a single FW-190 which succeeded in knocking out the number three engine, forcing the ship down toward the undercast." The bomb bays were on fire from leaking gasoline. Lt. Capalbo handed Lt. Rundbaken a screwdriver so he could loosen the throttle stops for more speed. Cpl. John A. Villars was on the flight deck setting the radar interpreter when the fighters attacked. He was either mortally wounded or knocked unconscious. Lt. Rundbaken, with no oxygen mask and in spite of the flames, opened the bomb bay doors. He was severely burned and was overcome by the heat and lack of oxygen. Lt. Rundbaken collapsed on the catwalk of the forward bomb bay. Lt. Capalbo threw Lt. Rundbaken out the bomb bay, but his chute was too severely burned to have saved him. The nose gunner, Cpl. Richard McGuire, was out of his turret and on the flight deck with the radio man when the bomber exploded. Lt. Rundbaken, Cpl. McGuire, and Cpl. Villars died in this action.

AAF #	Type	Group	Sq.	Sq. #	Nickname
42-28913	B-24H	461st	767th	67	Unknown

PILOT: 1st Lt. Robert A. Galvin, San Antonio, TX, Single
MISSIONS: Unknown
LAST SIGHTED: 12:00, Muglitz 49-47′ N 16-56′ E
WHERE: Crash landed – Rosszyne, Hungary

#67 was flying in the #2 position. Lead flight navigator 2nd Lt. Frederick C. Smith said, "we were ten minutes, 20 miles south of Odertal when fighters hit us." S/Sgt. Robert F. Miller, tail gunner on the lead ship, "67 took a hit in the #3 engine from an FW-190. The #3 engine burst into flames and the bomber pulled off and dropped away." Sgt. Piccoli, though wounded, smothered the flames in the tail section with his body. The entire crew returned to the states.

AAF #	Type	Group	Sq.	Sq. #	Nickname
42-51322	B-24J	461st	767th	66	Unknown

PILOT: 1st Lt. Charles V. Lang, Jr., Carrollton, AL, Married
MISSIONS: Unknown
LAST SIGHTED: 12:00, Muglitz 49 – 47′ N 16 – 56′ E **DOWN:** 12:30
WHERE: Rytaik, Rybnik, Skrzischow. Guy Bosso says he bailed out over Adony, Hungary

The aircraft left formation 150 miles NW of Budapest. The navigator, Arthur W. Mohlman, left formation 50 miles north of Lake Ballaton. S/Sgt. Robert F. Miller, the tail gunner in the lead ship, "#66 was 600-700 yards behind the formation at the time of the first attack." T/Sgt. H.N. Ryan in ship #60, "#66 went under our aircraft. All engines were going and no fighters were after him - - then an FW-190 went down after #66." T/Sgt. Russell H. Ritchel, right waist gunner aboard #63 in number 2 position in C Flight, first section (767th), #66 pulled up alongside of us from the rear with a fighter on his tail. Fire and smoke were coming from the #3 engine." Cpl. Guy V. Bosso, right waist gunner said Lt. Lang ordered him to check on a fuel leak in the bomb bays. #66 peeled off to the right and fell away. The enemy fighter on his tail burst into flames. #66 fell about 500 yards, blew up and fell in two pieces. The entire crew got out and 7 were captured about 3:45PM. German reports note, "Cpl. James M. Rutter "refuses to talk!" Crew members will not even reveal the type of aircraft they were flying. Lt. Lang and three others were picked up by the Russians and returned to the 15th Air Force.

AAF #	**Type**	**Group**	**Sq.**	**Sq. #**	**Nickname**
44-41158	B-24J	461st	767th	77	Unknown

PILOT: 2nd Lt. Max L. Hailey, KIA, Walnut Grove, MO, Married
MISSIONS: Unknown (Sgt. Jacob Herman III had 49 missions)
LAST SIGHTED: 12:00, Muglitz 49-47′ N 56-16′ E **DOWN:** Unknown
WHERE: Pressnitz near Bruenn

#77 was flying number 3 position in the first attack unit. S/Sgt. Robert F. Miller, tail gunner, lead ship in formation, "#77 was way out in back . . . 1,000 yards behind. Fighters made repeated attacks on both planes, #66 and #77 until #77 was finally hit, peeled off and headed down." Hailey was downed by an FW-190 per Arthur G. Hermansen, Jr., left waist gunner on the lead ship. Sgt. Albert Thomas Arrotta, ball gunner on #77, "Fighters knocked out three engines." The bomber fell through the undercast in flames. A crew member yelled to Lt. Hailey to "Feather the prop!" There was no answer from Lt. Hailey. 6 KIA.

AAF #	Type	Group	Sq.	Sq. #	Nickname
42-51324	B-24J	461st	764th	12	Ten Men Bak/Paulette

PILOT: 2nd Lt. Kenneth B. Smith, Indianapolis, In, Married
MISSIONS: Unknown
LAST SIGHT: One hour from target
WHERE: 50 10' N 17 00' E

Lt. Smith called Capt. Marion Mixson, CO of the 764th Bomb Squadron, to inform him that he had one turbo out, one rough engine, and was low on fuel. Capt. Mixson said he should stay with the formation if possible. He later left the formation without making contact with the lead ship. He either ran out of fuel and bailed out his crew, or was attacked by fighters. James G. Taylor, 764th pilot said, "I was flying in the #5 position of "A" flight and Lt. Smith was in #7 position of the same box. Lt. Smith called Capt. Mixson to say that he had a bad engine and needed to abort. I later heard he may have joined "A" flight, second attack unit and went over the target. 1st Lt. Clark Barrett said "My tail gunner said #12 was dropping out of formation. I noticed a 461st ship while under fighter attack, which was at 5 o'clock low being attacked by 2 or 3 German FW-190s". Nine of the crew became POWs while ball turret gunner Cpl. Roland Morin was killed in the action.

AAF #	Type	Group	Sq.	Sq. #	Nickname
42-51430	B-24J	461st	765th	24	Tulsamerican

PILOT: 1st Lt. Eugene P. Ford, Blairsville, PA, Married, KIA
MISSIONS: Unknown
LAST SIGHTED: Off the coast of the Isle of Vis, Yugoslavia

While enroute to the target Lt. Ford's ship was heavily damaged by fighter attack, forcing him to jettison the bombs and attempt a return to Torretta. The Tulsamerican had lost one engine, hydraulics and was leaking fuel. The bomb bay doors were also stuck open. While over Hungary they encountered flak batteries and sustained more damage. Realizing they could not make it to Italy, Ford decided to attempt a landing at the emergency airfield on the Isle of Vis. Because they could not extend the nose gear, Lt. Ford was forced to go around for another attempt. However, before he could reach the runway the ship ran out of fuel forcing him to ditch in the Adriatic Sea, hitting the water at approximately 150 miles per hour. The plane hit, flipped, and broke in half. Lt. Ford, Navigator 1st Lt. Russell C. Landry, and Engineer / Right Waist Gunner T/Sgt Charles E. Priest were unable to escape the ship before it sank. S/Sgt. Edward "Eddie" Steelandt, flying as a Left Waist Gunner replacement on the mission, grabbed onto his fifty-caliber gun in the waist as the plane went in. He exited through the waist gun position by going under his gun which was now above his head in the upside down fuselage. Not seeing the tail gunner, Sgt. Steelandt reentered the sinking aircraft and rescued S/Sgt James Hazel, pulling him from the wreckage. In all, seven men were able to escape the aircraft and were picked up by fishermen and a British rescue team about 45 minutes after the crash. The Tulsamerican was the last B-24 manufactured at the Tulsa, OK plant.

AAF #	**Type**	**Group**	**Sq.**	**Sq. #**	**Nickname**
42-50934	B-24J	484th	824th	28	Little Joe

PILOT: 1st Lt. Charles M. Himmler, Trafford, PA, Single, KIA
MISSIONS: 13-14
LAST SIGHTED: According to MACR 1252, Spombataly, Hungary. German records show it shot down at 12:00, fighter attack at 11:45 with direct hits on the bomb bay and waist.
WHERE: Disappeared between Vienna and Lake Balaton. German records indicate Wenzelsdorf, CZ 14 miles NW of Sternberg.

#28 was flying Baker 12 position. It lagged behind for quite some time. The #2 engine was overheating. At the time enemy fighters hit the 461st Bomb Group, Lt. Himmler called on VHF and told Lt. Olson that he was going to abort. Lt. Himmler dropped his bombs just as two enemy fighters made a pass at him. #28 reported 6 fighters at 6 o'clock high, 2nd Lt. Howard C. Long, navigator, said "Give 'em hell" over the interphone as the fighters attacked. The bomber took hits in the waist and bomb bay. The B-24 burst into flames, went out of control, and entered a spin. Seven crew members and Sgt. David V. Uber, aerial photographer went in with the plane. S/Sgt. Frank W. Carr, the nose gunner, was hidden by civilians for 10 days before he was turned over to the Germans.

AAF #	Type	Group	Sq.	Sq, #	Nickname
42-51941	B-24J	451st	726th	47 0	Unknown

PILOT: 1st Lt. Theodore C. King, Brookline, MA, Single, KIA
MISSIONS: Unknown   Deputy Leader
LAST SIGHTED: Seen to crash at 12:55
WHERE: 50-28" N 18-04" E, Ringwitz: 5km NE of Friedland

#47 was flying in the number 2 position as Deputy Leader. Pally after target was to left. The lead (Shelton) veered off to the right. Lt. King tried to fly over him and take over the lead.

Capt. Shelton pulled back into position, his prop cutting off the tail of Lt. King's B-24, which entered a flat spin. All 10 crew members went in with their ship. The only survivor was S/Sgt. Joseph Weisler, the photographer, who was captured at 1:30PM near Ringlitz Forest.

AAF #	Type	Group	Sq.	Sq. #	Nickname
42-52045	B-24J	451st	726th	54 A	Unknown

PILOT: Capt. William T. Shelton, Janesboro, AK, Married
MISSION: Unknown, Flight Leader
LAST SIGHTED: 50-31' N 18-08' E
Down: 16:30
WHERE: Plevja, Yugoslavia

Midair collision with Lt. King's aircraft 40 miles NW of target at 26,000 feet, weather clear, ground snow covered. The nose gunner, navigator and bombardier bailed out at 13:00. Two were captured at Floste at 13:00. However, Lt. John L. Sullivan, Jr., the bombardier, was

KIA. The aircraft stayed in the air until 16:30 and crashed 18 miles from Plecia. T/Sgt R.J. Bolger stated that bailout was at 5:30.

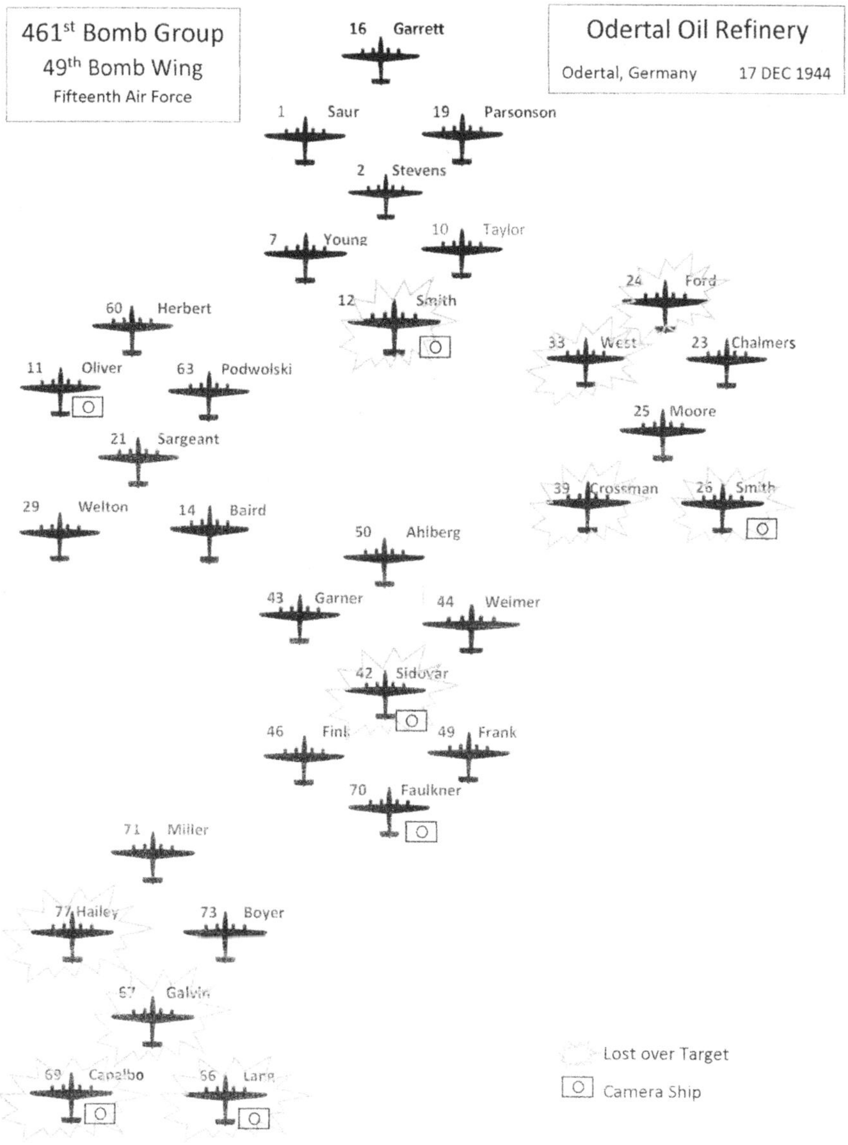

SUPPLY MISSIONS:
May 1945

COMBAT MISSIONS were at an end, but there were still missions flown in May 1945. The 461st was given the responsibility of flying food and medicine to the Spittal and Wolfsburg POW camps in southern Austria. American and other allied personnel had been abandoned without desperately needed supplies.

Between 9 May 1945 and 15 May 1945, nine mission were flown in what veterans described as "the most satisfying missions of the war". 461st The Liberaiders dropped 184 tons of food and medicine to maintain the former POWs until ground forces were able to reach them.

1 Spittal, AUT - POW Supply Mission 9/5/1945

28 461st aircraft dropped supplies on the Spittal POW camp. Two of those ships dropped 12 containers of medical supplies while the rest dropped 12 containers of urgently needed food. Some containers landed in the camp and some in the surrounding fields. One chute failed to open and the container fell through the roof of one of the barracks. A large number of inmates rushed for the containers both in and outside the camp. Some civilians were also seen trying to collect food. A total of 336 containers were dropped.

2 Wolfsburg, AUT - POW Supply Mission 10/5/1945

Panels would be displayed at the target, but the drop will be accomplished even if the panels were not present. 24 aircraft successfully dropped food and supplies on the Wolfsburg POW camp. As on the last mission the lead ship remained over the target and acted as control ship for the three aircraft elements which dropped in eight waves. Most containers were dropped west of the camp, which was considered better suited for dropping of supplies. Each aircraft dropped 12 containers of rations. Crews were to carefully observe any activity in the area of the camp

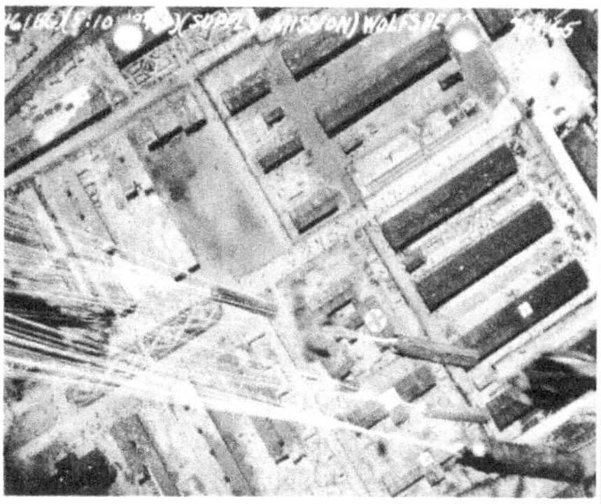

.

3 Spittal/Wolfsburg, AUT - POW Supply Mission 11/5/1945

Supplies for the days mission were delivered by the 125 D.I.D. (British) from Foggia and will continue each day until cancelled by 49th Bomb Wing Headquarters. The 461st Bomb Group dropped 3,000 rations on both the Spittal and Wolfsburg POW camps in clear areas of the camps. The Group used free dropping methods and dropped all containers within the target area marked by panels. There was some difficulty dropping the containers due to rack malfunction, but all supplies were successfully dropped.

4 Spittal/Wolfsburg, AUT - POW Supply Mission 12/5/1945

Blue Force, consisting of four aircraft dropped supplies on the Wolfsburg POW camp. Red Force, also consisting of four aircraft, dropped supplies on the Spittal POW camp. All aircraft dropped supplies using the free fall method. German troops were observed moving North of Wolfsburg.

5 Spittal/Wolfsburg, AUT - POW Supply Mission 13/5/1945

Four 461st B-24s each dropped 6,000 rations on the Spittal POW camp and another four B-24s dropped 6,000 rations on the Wolfsburg POW camps. All rations were dropped by free fall method successfully.

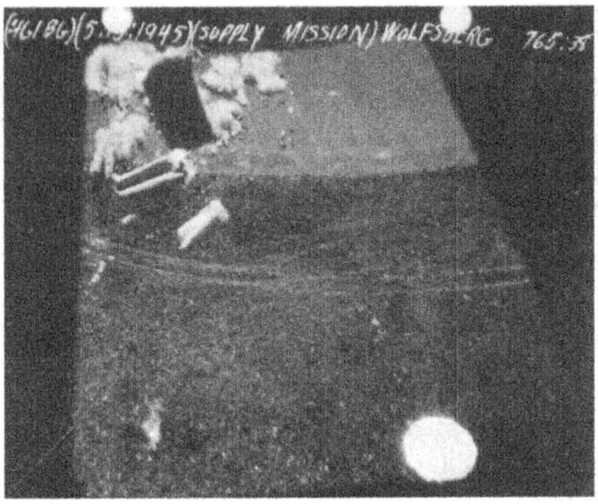

6 Spittal/Wolfsburg, AUT - POW Supply Mission 14/5/1945

Four aircraft dropped 2,000 rations each on the Spittal POW camp and another four aircraft dropped 2,000 rations each on the Wolfsburg POW camp. Drop was made successfully using the free drop method. C-47s and Spitfires were observed on the ground at Klagenfurt. Large supply convoys were seen heading South and West. No German troop activity was observed.

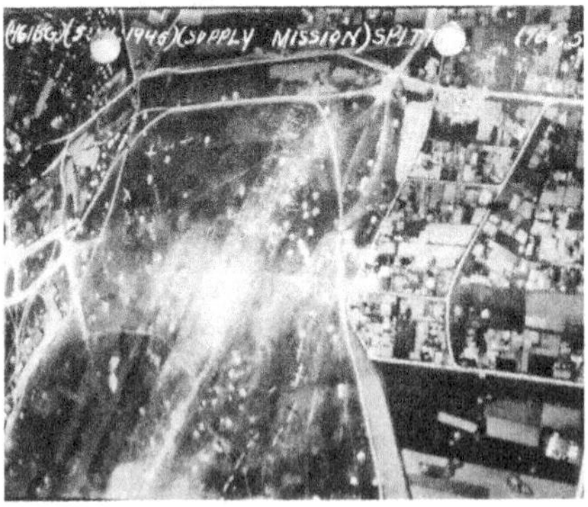

7 Spittal/Wolfsburg, AUT - POW Supply Mission 15/5/1945

Four aircraft each dropped 1,640 rations on the Spittal POW camp while another four aircraft dropped 1,640 rations on the Wolfsburg POW camp. All rations were dropped by free fall successfully. A large truck convoy was observed coming into Celje from the North West and another large truck convoy was observed moving into Spittal from the South East. This would mark the last mission flown by the 461st Bomb Group. They had successfully flown desperately needed food and medicine to Allied prisoners until truck convoys were able to reach the area.

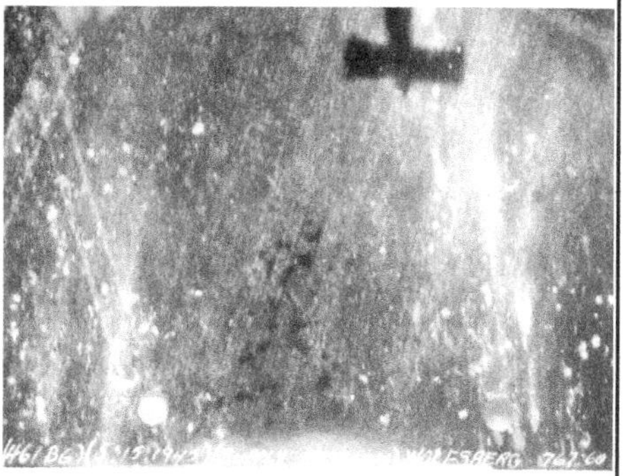

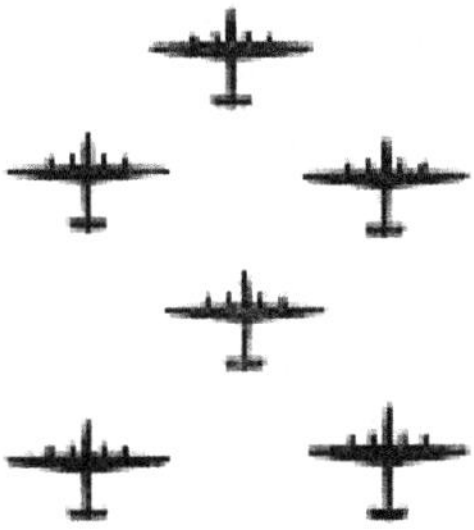

(5)

OTHERS

XV AIR FORCE SERVICE COMMAND

81ST AIR SERVICE SQUADRON

ARMY AIRWAYS COMMUNICATIONS SERVICE

ORGANIZATION OF THE XV AIR FORCE SERVICE COMMAND

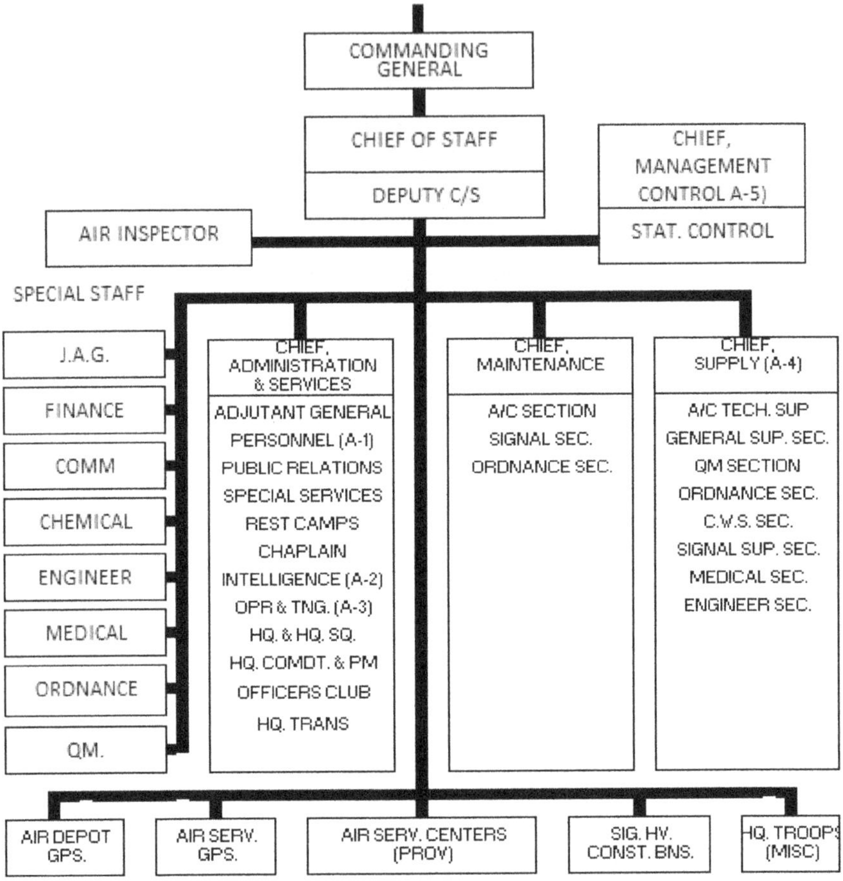

FIFTEENTH AIR FORCE SERVICE COMMAND

FIFTEENTH AIR FORCE SERVICE COMMAND (XV AFSC) was just what it sounds like, a service organization. While volumes have been written about the tactical units of the Fifteenth Air Force, relatively little is known about the service units. In fact, it was the tactical units and the service units working together that provided the Fifteenth with the ability to aggressively pursue the war efforts and bring it to a successful conclusion. "The operational requirements of the Fifteenth Air Force governed very largely, of course, the operations of the XV Air Force Service Command. As far as was possible under the rapidly changing demands of air warfare in the Mediterranean Theater, the headquarters of the two organizations, working together in close liaison, tried to anticipate those requirements. It was the responsibility of the Service Command headquarters, however, to see that the requirements were met—to establish policies and plans that would ensure that adequate supplies, equipment, and facilities would be available at the right place and time. The actual task of providing supply and maintenance service to the combat groups rested most directly with the field units of the Service Command, but the plans and policies established by the headquarters set the pattern of operation of the field units." [1]

XII AFSC had been the organization responsible for supplying every unit in the Mediterranean Theater of Operations from Casablanca to Algiers. However, it became impossible to control such a vast territory from one location. To solve the problem XII AFSC created three area service commands while keeping control over all Service Command functions.

---

(1) History of the Fifteenth Air Force, Volume 1, Part I: Composition of the Air Force, Chapter 2, page 83

XII AIR FORCE SERVICE COMMAND

I ASAC (Air Service Area Command)
II ASAC
III ASAC

The creation of the Fifteenth Air Force Service Command really began on 22 December 1943, when NATOUSA issued an Advanced Directive which completely reorganized the Air Force Commands in the Mediterranean, both operational units of the Air Force and Air Service personnel. The creation of the Fifteenth Air Force necessitated wholesale changes to XII Air Force Service Command and its subcommands. It also created some territorial challenges that can make the various roles of each organization difficult to understand. XII AFSC became Army Air Forces Service Command/Mediterranean Theater of Operations (AAFSC/MTO). One of the three subcommands, III ASAC was redesignated XII AFSC to support XII Air Force.

The next step in the evolution of the organization would occur on 1 January 1944 when Army Air Forces/Mediterranean Theater of Operations issued General Order No. 1 disbanding II Air Service Area Command and redesignating it XV Air Force Service Command (Hq., AAF/MTO GO No 1, 1 January 1944). The next day XV AFSC would be officially assigned to Fifteenth Air Force (Hq., AAF/MTO GO No 2, 2 January 2, 1944) and made retroactive to 1 January 1944.

The same General Order that created Fifteenth Air Force Service Command also added a number of units including Headquarters and Headquarters Squadrons, Air Depot Groups, Service Groups, and a list of repair, supply, and maintenance as well as signal, ordnance, and engineering units.

ORGANIZATION EFFECTIVE 1/1/1944

XII AFSC becomes AAFSC/MTO and included:
Adriatic Depot (SP)
I Allied Air Force Area Command (SP)
II ASAC becomes XV AFSC
III ASAC becomes XII AFSC

The next step for XV AFSC was a general order establishing service centers and depots. (Hq., XV AFSC GO No. 3, 8 January 1944) While the general order was published on 8 January, it was also retroactive to 1 January 1944. Section I of General Order 3 would create the following

Service Centers:

- Service Center No. 37 serving elements of the 55th Bomb Wing, 15th A.F. (upon arrival)
- Headquarters; Foggia, Italy
- Service Center No. 38 would serve elements of the 49th Bomb Wing, 15th A.F. upon arrival.
- (It would later be reassigned to provide services to the 305th Fighter Wing while the 49th Bomb Wing would receive services from the 37th Service Center)
- Headquarters: Lucera, Italy
- Service Center No. 43 served elements of the 304th Bomb Wing, 15th A.F. (upon arrival) Headquarters: Cerignola, Italy
- Service Center No. 62 would service elements of the 47th Bomb Group, 15th A.F. Headquarters: Manduria, Italy
- Service Center No. 324 To service elements of the 5th Bomb Wing, 15th A.F. Headquarters: Foggia, Italy

Section II established the following depots:

Air Force General Depot No. 5
Headquarters: Bari, Italy
Operated by 11th Air Depot Group

(Later transferred under AAFSC/MTO. After the transition, Army Air Force supplies at AFGD #5 would be managed by the 11th, 22nd, 32nd, 41st, and 85th Supply Squadrons. By the end of the war they would be overseeing approximately half the supplies of all AAFSC/MTO depots.

Repair work was the responsibility of the 11th, 31st, 85th and 90th Repair Squadrons, who completed about 1800 work orders per month. Some of the work involved repairing equipment that was placed back in inventory. However, considerable work was done on B-24s forced down in Yugoslavia and brought to Bari for repairs.)

Air Force Advanced Depot No. 51
Operated by 41st Air Depot Group
Headquarters: Oria, Italy

Air Force Advanced Depot No. 52
Operated by 41st Air Depot Group
Headquarters: Gioia, Italy

Air Force Advanced Depot No. 53
Operated by 18th Air Depot Group
Headquarters: Foggia, Italy

By the end of the first month of operation there were a total of 157 service and depot units of all types totaling 837 officers and 14,662 enlisted personnel.

In early February, an order was signed announcing the establishment of the following chemical ammunition depots:

Air Force Chemical Ammunition Depot No. 5
Headquarters: Orta Nova

Intransit Air Force Chemical Ammunition Depot No. 5A
Headquarters: Valenzano

Air Force Chemical Ammunition Depot No. 7
Headquarters: Latieno

Another part of that order established two other ordinance depots within Fifteenth Air Force Service Command.

Air Force Ordnance Depot No. 8
Headquarters: Manfredonia

Air Force Ordnance Depot No. 11
Headquarters: Francavilla

One month later, all chemical ammunition depots and ordnance depots would join General Depot 5 with assignments to Army Air Forces, Mediterranean Theater of Operations.

To better understand XV AFSC, and its role in the air war during 1944 – 1945, it is important to discuss both the organization and the operation of these service units.

To perform their duties adequately XV AFSC would have an extensive list of both major and secondary responsibilities. Major responsibilities included:

- Third and fourth echelon technical air corps supply,
- Third and fourth echelon aircraft maintenance and repair,
- Third and fourth echelon supply and maintenance of all motor vehicles assigned to the air force,
- And supply and maintenance of all ordnance and chemical warfare bombs and ammunition.

While a complete list of all secondary responsibilities would be impossible, the more important ones consisted of:

- Establish telephone and telegraph facilities where none had existed,
- Repair and maintain existing facilities and operate those facilities,
- Furnish certain kinds of quartermaster equipment and supplies not available elsewhere,
- Supply trained personnel to perform all service command functions,
- Operate motor transport services,
- Provide financial services, and
- Maintain certain types of operations and flight, engineering, photographic, and medical services.

To provide a basic understanding of echelons of maintenance during World War II, the Army Air Forces used four levels, or echelons, depending on what unit would perform maintenance services.

- First echelon maintenance consisted of servicing aircraft and equipment, preflight and daily inspections as well as minor repairs, adjustments, and replacements which were performed by the air echelon of the combat group.
- Second echelon maintenance involved servicing aircraft and equipment, preventive inspections. This also includes engine changes under certain circumstances. Second level maintenance was usually performed by ground units of the combat unit, air base squadrons, or airways detachments.
- Third level maintenance, requiring personnel with specialized training, involved field repairs and salvage, removal and replacement of major unit assemblies, fabrication of minor parts, and minor repairs to aircraft structures and equipment. Services were performed by service groups and sub-depots.
- Fourth echelon maintenance required all operations necessary to restore aircraft to combat status and major overhaul of engines and other assemblies or accessories, and the recovery, reclamation, or repair and return to service of aircraft incapable of flight. Fourth echelon maintenance was the responsibility of air depot groups and air depots.

* * * * *

GENERAL DEPOTS

As was mentioned above, Air Force General Depot #5 was transferred to Army Air Forces Service Command/Mediterranean Theater of Operations in March 1944 largely because it could function as a fourth-echelon supply installation capable of carrying supplies for greater than 90 days which could provide for the needs of every Army Air Force unit in Italy. Because it was closer than any advanced depots, it also functioned as an advanced depot for the service center at Spinazola.

The operation of the maintenance function was also not impacted by the transfer to AAFSC/MTO, largely because the Chief of Maintenance at XV AFSC was also Chief of Maintenance at Depot #5. Like the advanced depots, it functioned as a fourth-echelon maintenance facility although each depot developed a certain specialization. For Depot #5 that meant responsibility for the repair of all fighter aircraft in the Air Force, and for major repair of salvaged fighters. They also kept a large stock pile of repairable parts from which any unit could draw from reducing the number of aircraft grounded for lack of parts.

Depot #5 acted as a proving ground for modifications suggested by other units, or when directed by technical orders. If a modification were accepted they could manufacture the necessary parts (or have the advanced depots manufacture them) and distribute technical orders.

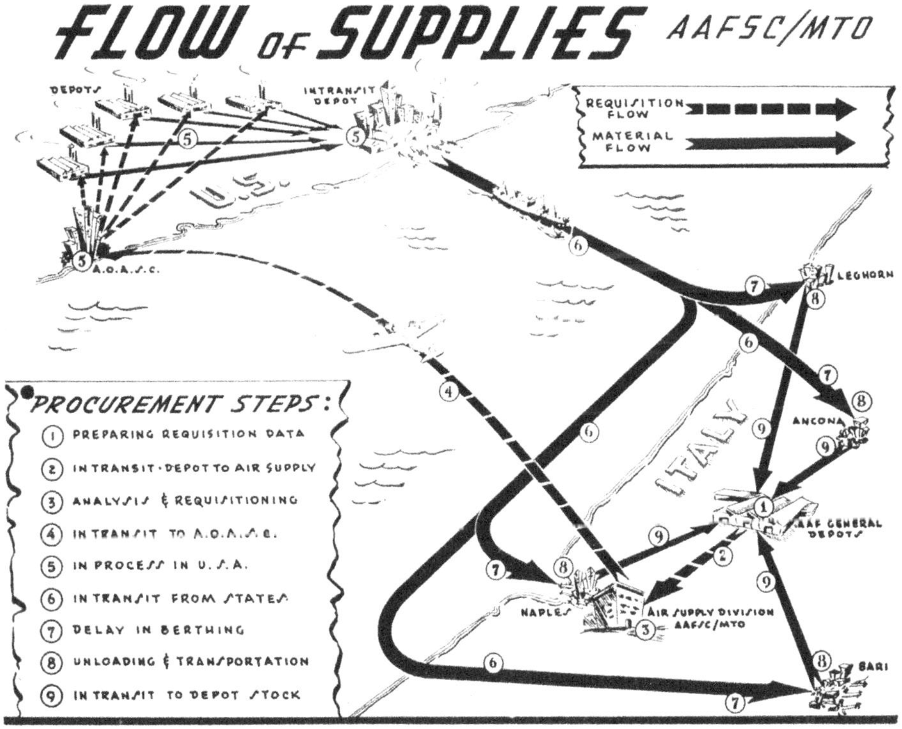

MEDITERRANEAN ALLIED AIR FORCE SUPPLY CHAIN

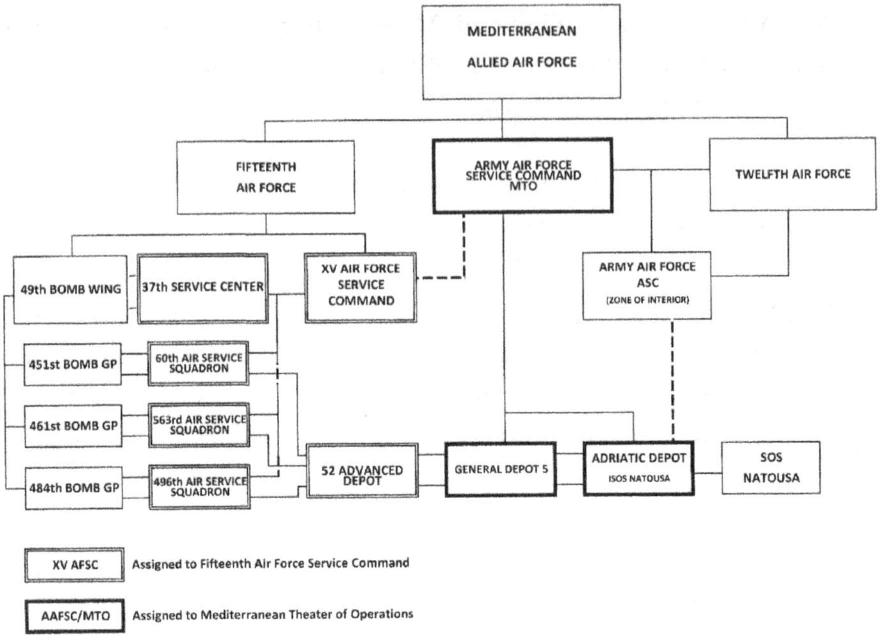

ADRIATIC DEPOT

21 October 1943 - Adriatic Depot was established in the Bari - Foggia area and placed under the control of, and attached to, AAF/MTO through AAFSC/MTO. The depot established all the services of a full Base Section, including transporting supplies to forward areas.

Adriatic Depot included:

	COVERED	OPEN	TOTAL
Depot No. 1 (Bari)	240,147	46,505,117	46,745,264
Depot No. 2 (Foggia)	374,656	3,100,470	3,475,126
Depot No. 3 (Ancona)	9,300	1,166,500	1,175,800

*** The 15th Air Force flew supplies of 100-octane gasoline from the Foggia pipeline to airfields in France to support the rapid advance of the Seventh Army up the Rhone Valley, using drums supplied by the Adriatic Depot.***

* * * * *

ADVANCED DEPOTS

Under the 41st Air Depot Group was the 51st Advanced Depot at Oria and the 52nd Advanced Depot at Gioia. Advanced Depot 51 was exclusively a supply installation providing fourth echelon supplies for the 47th Bomb Wing and 62nd Service Center. It included a depot supply squadron and a quartermaster truck company. Advanced Depot 52 was a maintenance facility providing fourth echelon work exclusively for B-24s. When at full strength Depot 52 supported 15 bomb groups from the 47th, 49th, 55th, and 304th Bomb Wings.

- **51st Advanced Depot (Oria)**
  - 87th Depot Supply Squadron
  - 2460th Quartermaster Truck Company (AVN)
  - Oxygen Detachment (Attached to 87th DSS)

- **52nd Advanced Depot (Gioia)**
  - Hq. & Hq. Squadron
  - 41st Depot Repair Squadron
  - 88th Depot Repair Squadron
  - 312st Depot Repair Squadron
  - 1044th Signal Company (SG)
  - 2500th Quartermaster Truck Company (AVN)
  - 432nd Quartermaster Platoon (Attached)
  - 2075th Engineer Battalion (Attached)
  - 41st ADG Combat Crew Reception Center (Attached)

The 53rd Advanced Depot at Foggia, operated by the 18th Air Depot Group, included both supply and maintenance capabilities providing services only to the B-17s of the 5th Bomb Wing.

- **53rd Advanced Depot**
  - Hq. & Hq. Squadron
  - 84th Depot Repair Squadron
  - 87th Depot Repair Squadron
  - 319th Depot Repair Squadron
  - 31st Depot Supply Squadron
  - 469th Quartermaster Platoon Advanced Group
  - 1956th Quartermaster (Truck) Company (AVN) c
  - 1964th Quartermaster (Truck) Company (AVN) c
  - 2467th Quartermaster (Truck) Company (AVN)
  - 875th Signal Company Depot (AVN)
  - Det. 2091st Engineer Aviation F/F Platoon
  - 14th Medical Supply Platoon

"The fourth echelon of maintenance performed by the Depot Repair Squadrons assigned to Depots 52 and 53 included all operations necessary to completely restore worn or damaged aircraft to tactical serviceability; the periodic major overhaul of engines, unit assemblies, and auxiliary equipment; the fabrication of parts required in an emergency or directed by technical instructions; the accomplishment of technical compliance changes as directed; and the replacement, repair, and checking of auxiliary equipment. Another function of the repair squadrons was the fourth echelon of reclamation of aircraft, which included the disassembly, repair, and return to stock of component parts, or the salvage or destruction of material damaged beyond use or economical repair."

* * * * *

SERVICE CENTERS

Air Service Centers, or Air Service Groups, were commanded by a Colonel or Lt. Colonel with assistance from a Chief of Staff or Executive Officer, and supported by the Headquarters and Headquarters Squadron. Officers commanding the various service squadrons assigned to the Service Center functioned as Staff Officers.

A separate Service Group provided 3rd echelon support to each Bomb Wing (The 37th Air Service Group supported the 49th Wing). Since the duties performed by a service group had a direct bearing on tactical and strategic operations, control over them transferred to the XII and XV Air Forces during the reorganization. Each Service Group had Air Service Squadrons and detachments assigned directly with the individual groups. These service centers were mobile organizations provided to establish and operate third echelon maintenance and supply points within close supporting distance of the combat unit.

The 37th Service Group, as it relates to the 461st Bomb Group, consisted of:

ORGANIZATIONAL CHART
SERVICE SQUADRON
HEAVY BOMBARDMENT

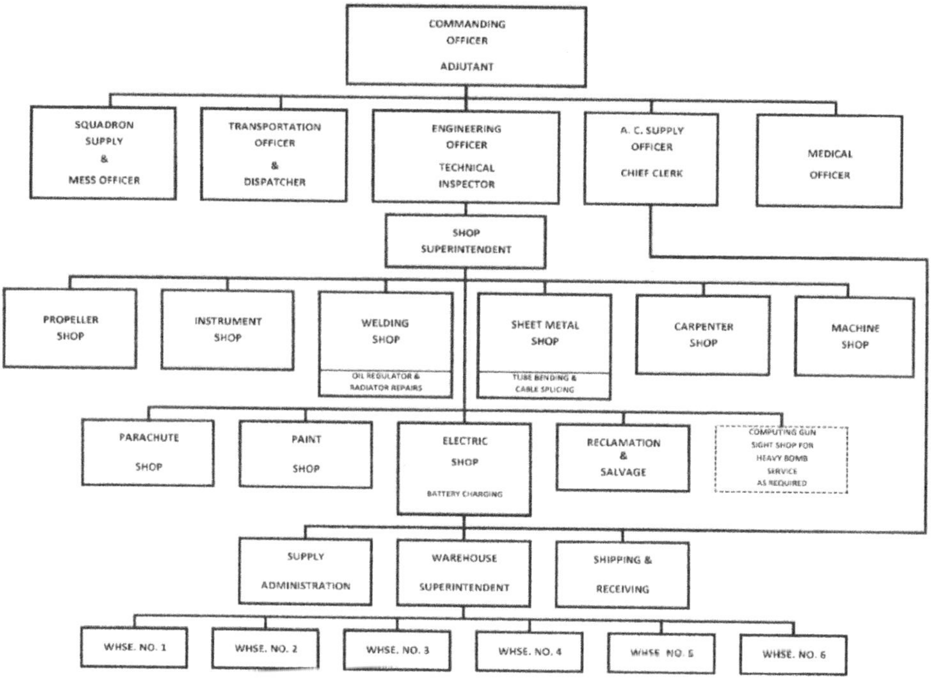

Aerial view of the 563rd Air Service Squadron area at Torretta Field taken in early May 1945. The long, black taxiway and hardstands belong to the 764th Bomb Squadron of the 461st Bomb Group. At the top of the taxiway, on a 764th hardstand, is a B-17 currently assigned to the 764th as a squadron hack. This ship is probably B-17F-55-BO 42-29527, nicknamed "Jig-Jig". It had been transferred from the 483rd Bomb Group to Fifteenth Air Force Service Command and later transferred to 49th Bomb Wing.

All repair work was performed on the large, black area currently occupied by two B-24s and a number of smaller aircraft, possibly P-40 fighters, and L-5 observation aircraft. A third B-24 is waiting its turn and a B-17 is toward the top of the work area. The center of the photo shows the framework of a hanger being constructed to provide shelter during inclement weather. Unfortunately, it was never completed. At the bottom of the picture is the aircraft salvage yard. All usable parts were salvaged before the fuselages were dragged to the "bone yard" for disposal.

*   *   *   *   *

- The 563rd Air Service Squadron, primarily responsible for third echelon repair or salvage of 461st Bomb Group aircraft and included:
    - Propeller Shop – Remove, balance, clean & install. Also service on ship
    - Instrument Shop – Check, repair, clean, install, or turn into supply
    - Electric Shop – Charge batteries, repair generators, repair/modify power plants
    - Parachute Shop – Inspect, repair, replace parachutes, modify life vests.
    - Paint, Dope & Fabric Shop – Repair battle damage, paint cowl rings
    - Airplane Mechanic & EM Shop - Change, install, pickle engines
    - Sheet Metal Shop - Repaired battle damage, repaired bomb bay doors
    - Armament Shop – Salvage and repair battle damaged weapons
    - Carpenter Shop – Manufacture floor boards, spark plug racks, engine cover plates, etc.
    - Machine Shops – Manufacture bolts, bushings, rivet covers cleaning rods
    - Welding Shop – Cut up salvaged ships, manufacture ammo chutes.

Salvage operations on a Liberaider that crash landed at Torretta Field

Engine repairs on #41 of the 766th Bomb Squadron, March 1945

Painting the distinctive red bar on a new 461st aircraft

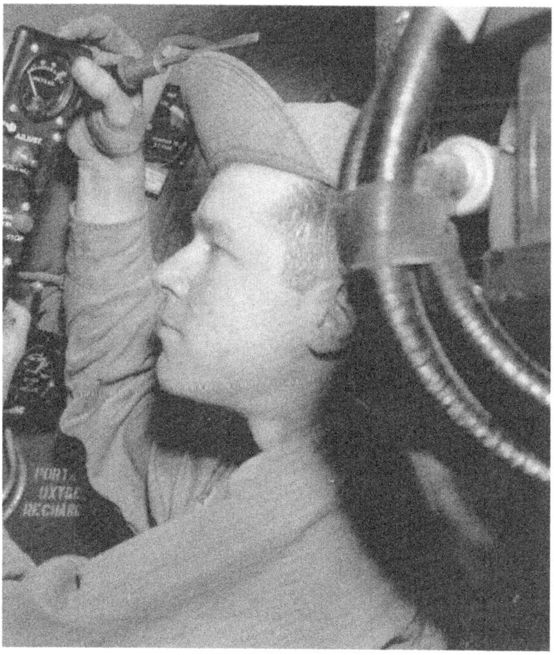

Dual installation of a radar jammer. The majority of 461st ships were equipped with this set

563rd Air Service Squadron engine salvage

461st Bomb Group salvage yard. The 563rd would drag stripped fuselages here after removing all useable parts

One other important service provided by the 563rd Air Service Squadron was the salvage of aircraft no longer considered to be economically repairable. The policy, titled "Disposition of Unserviceable Aircraft," which was published by Headquarters, Army Air Force Service Command/Mediterranean Theater of Operations, laid out the process involved in salvaging an aircraft and the individuals responsible for making sure it ran smoothly. Headquarters, Fifteenth Air Force Service Command also published a memorandum outlining the process. When an aircraft became damaged beyond economical repair in the judgement of an authorized Engineering Officer of the Service Squadron, the Salvage Officer of that Service Squadron would assume responsibility for that aircraft. A guard would be placed on the aircraft who would be on duty until the salvage operation was completed. The salvage team would disassemble the ship in a specific order:

- First priority was the removal of AOC items or items that may be damaged if left on the aircraft.
- Next would be removal of propellers, engines, automatic flight controls, bombsight, and weapons. Critical items needed to repair other aircraft waiting for those items to become operational were removed in the order that they were needed. XV Air Force Service Command was responsible for sending Salvage Officers a daily report to assist them in prioritizing and shipping items with the highest need.
- Finally the disassembly of major sections of the aircraft would come.

The Depot Reclamation and Salvage Officer was responsible for informing the Depot Supply Officer of the location of salvaged aircraft to assist them in obtaining needed parts. The former policy of reducing remaining parts to scrap was discontinued at this point. There was also a decision whether or not to leave the ship where it crashed or bring it back to the disposal area. Any aircraft abandoned at the scene of the crash must in accordance with AAF Regulation No. 65-85, Section VI.

* * * * *

A Private from the Signal Company keeps the lines of communications alive by repairing this teletype machine at Torretta in February 1945.

The 1153rd Signal Company began servicing the 461st Bomb Group on its arrival February 27, 1944. It would be responsible for setting up courier services between the 37th Service Group, 461st Bomb Group, and HQ, Fifteenth Air Force, running a radio repair section (B-24), maintaining a radio repair truck, hauling gravel, and compacting roads, and laying spiral four cable and carrier wire.

* * * * *

1434th Ordnance Supply & Maintenance was responsible for servicing two bomb groups (461st and 484th), two service squadrons, one truck company, one air base security company, one signal company, and one quartermaster detachment. Their general responsibilities included:

- Automotive – General checkup and repair, welding radiators, valve jobs, body and fender repairs, machine shop, engine overhaul, transmission, and rear end jobs,
- Ammunition – Linking .50 caliber ammunition, and supplying bombs,
- Armament – Inspection, cleaning & repairing of small arms,
- General Supply – Automotive parts salvaged and secured for 16 vehicles.

1153rd linemen at work. The building houses the switchboard, radio repair detachment, and signal supply detachment

1434th Ordnance loading bombs at the Torretta bomb dump

The 1434th prepares another truck load of 500 lb. bombs for the flight line in preparation for the days mission.

Acetylene welding by a member of the 1434th Ordnance S&M Company, January 1945

* * * * *

- The 2239th QM Truck Company operated the motor transport pool and provided motor transport service, including:
    - Providing third echelon motor vehicle maintenance and operating class I and class III refilling points,
    - Providing quartermaster salvage service,
    - Procuring, storing, and issuing quartermaster class II and IV supplies for air force units,
    - Providing purchasing and contracting services.

2239th Quartermaster Truck Company Motor Maintenance Section, February 1945

* * * * *

- Company 'H,' 3rd Battalion, 21st Engineer Aviation Regiment started work on Torretta Field late in November 1943 and remained at work with all platoons until they were relieved by Second Battalion, 21st Engineer Aviation Regiment on 1 June 1944. The Second Battalion built a gravel runway, 150' x 6000' and laid 500' of PSP at each end of the runway. They widened 17,000 linear feet of compacted gravel taxiway 15 feet and built 50 heavy bomber type hard stands. They later laid 6,000 pieces of PSP on 800' at the south end of the runway as an experiment at the insistence of the Air Corps. Unfortunately, the planks rutted up due to foundation failure after a few days of heavy bomber traffic. Because of soil conditions, the field posed serious drainage problems during heavy rain from January well into March 1944. It was found impractical to put down gravel during rain, as the material absorbed and held water. The runway was opened to a depth of 30 inches, scarified, aired, and allowed to dry out. Then gravel, originally 6 to 8 inches thick, was re-compacted. An extremely flat surface, with slopes of less than ½ percent., caused water to stand on the field until a drainage system could be installed. Access roads and the erection of semi-permanent huts and buildings were included in field details.

- On 29 May 1944 Company "F", 2nd Battalion, 21st Engineer Aviation Regiment, assumed responsibility of this airfield and started to erect a Butler Combat Hanger. PSP 100' x 1000' was laid on both ends of the west runway. Drainage ditches were dug and the runway and taxiway consistently maintained. On July 1944 maintenance was taken over by the 1898th Engineer Aviation Battalion. A detachment from Company 'D', 21st Engineer Aviation Regiment, was assigned to assist the 1898th Engineer Aviation Battalion by constructing an all-weather, gravel type 100' x 6000' runway on the site of the existing dry weather runway. On 21 December 1944, Company 'F', 2nd Battalion, 21st Engineer Aviation Regiment, again resumed responsibility for the field. During January and the first two weeks of March a new gravel lift was placed on the east runway to replace the original surface that had an excess of clay. Company 'F' maintained this field until the end of the war at which time salvage operations were started.

- Also under the control of the 37th Service Group were:
    - 85th Battalion of the 47th Regiment (British Eight Army, Light Anti-Aircraft Unit)
    - "C" Troop of the 62nd Regiment (British Eight Army, Heavy Anti-Aircraft Unit)
    - 3rd Company Airport Interior Guard (Italian)

81st Air Service Squadron

THE 81ST AIR SERVICE SQUADRON organized on 24 June 1942 at Foster Field in Victoria, Texas; net strength 9 officers and 208 enlisted personnel. They would be assigned to the 43rd Service Group, part of Fifteenth Air Force Service Command following a move from Oudna #1, Tunisia to Sam Giovanni, Italy on 23 May 1944.

The Fifteenth Air Force recognized the need to establish emergency airfields for aircraft returning from missions with battle damage, mechanical failure, or low fuel. Those fields would be along the Dalmatian coastline along the route taken to Romania and the Ploesti oil fields. Without airfields pilots would be forced to choose between landing in rough terrain, bailing out over German held territory, or risking ditching in the Adriatic Sea.

In addition to repairing damaged aircraft at Bari, the 81st Air Service Squadron would be responsible for efforts to establish and maintain the emergency airfield on the Yugoslav Isle of Vis. The island was under the control of General Tito's Partisan forces. Helped by the British, an airfield to support the Partisans was constructed in May 1944 and was used by the RAF 352nd (Yugoslav) Squadron with Spitfires and the RAF 351st (Yugoslav) Squadron flying Hurricanes. While both were based on Italian airfields, they used Vis as a forward base. The runway itself was made of compacted gravel with steel matting at both ends and had been bulldozed into a shallow valley surrounded by hills and small mountains. Landing required a steep approach, slow airspeed and only a little room to roll after touchdown. Beginning in July 1944, the initial group, consisting of one officer and twenty men, were placed on detached service from the 81st to establish the operation on Vis. Another officer and twenty men quickly followed them. By September, the contingent had grown to eight officers and 228 enlisted men. During October, November, and December both Bari and Vis would repair or salvage 30-35 aircraft per month.

By January 1945, Vis had been firmly established. To manage the increasing workload 15 more 81st men made the move to Vis, as did 20 others picked from the 350th, 564th, 339th, 368th, and 556th Air Service Squadrons. Landings at Vis jumped to 47 for the month. Beyond repairing aircraft, the men winterized vehicles, made kitchen units, built machine shops, parachute shops, and carpenter shops. They repaired flying suits, metal folders, junction box covers, steam kitchens, and stoves. Also, during January the 81st was reassigned from the 43rd Service Group to the direct control of Fifteenth Air Force Service Command in Bari, Italy.

Two views of the approach to the runway on the Isle of Vis

44-40269, Rough Rid'er (far left) of the 766th Squadron

The next three photos show Yugoslavian Partisans as they assist the 81st Air Service Squadron men while they repair engines on the Isle of Vis.

44-41064, "V Grand" of the 465th Bomb Group so named because it was the 5,000th Liberator built by Consolidated.

42-94928 of the 484th crashed attempting an emergency landing. Three of her crew were killed in the crash.

Lady Duzz, 44-41044, under repair following an emergency landing by Lt. Alexander of the 764th

In the meantime another secret project, called "Project B," was working its way to Yugoslavia. At the end of the month 20 men assigned to Vis were called to establish a new emergency airfield at Pkros Airfield, Zara, Yugoslavia. On February 4, 1945 men and equipment left Bari to establish the emergency field at Zara. Upon arrival they became the only American unit in Yugoslavia other than a small detachment of the American Mission. They were armed with carbines and anti-aircraft guns, and surrounded by German forces. However, other than a monthly visit by a German reconnaissance plane and the sound of artillery, the camp was established in three days. They had set up a hospital unit, transient area for crews remaining overnight, transient supply unit, operations office, transportation unit with heavy vehicles, and a radio-telephone transmission unit. The first emergency landing was made one week after their arrival on 15 February 1945 when Lt. W.B. Bonifield of the 98th Bomb Group, 344th Bomb Squadron landed his B-24 (42-52022) with the #3 and #4 engines out.

When an aircraft arrived they found taxiing crews, parking crews, crash-rescue crews, checkers, transportation crews and mechanics, medical staff, food, housing, and if necessary, a plane for the trip home.

The men of the 81st Air Service Squadron were not just dedicated to aircraft repair. They also extricated men from the wreckage of their ships. If crew members were killed in flight, or were too critically wounded to survive, the men of the 81st prepared the graves, performed funeral services, and conducted final burials. They performed these services 28 times. Such was the case for Charlie Andrews of the 376th Bomb Group and the crew of the B-24 "Hardway Ten" that crash landed on Vis on 22 August 1944. During the landing, the ship turned upside-down killing three of the ten men aboard.

Commemoration ceremony for British and American servicemen on the Isle of Vis

"The plane was a mess and we were all inside of it. I can still hear the sound of one engine running; it had no prop but was turning over and my first thought was "fire." We will always thank the good Lord for the absence of fire as we would all have been dead, but some people on the ground got us out finally. I found out a few hours after I was pulled out that two of my crew were dead and that one man was in very serious condition with broken ribs that had punctured his lungs."

B-24J-155-CO, 44-40330, 376th Bomb Group, 515th Bomb Squadron seen following crash landing on Isle of Vis 22 August 1944.

The establishment of Zara allowed longer bombing and strafing missions, especially to more northern targets, for the Fifteenth Air Force. At the same time it would mean a decrease in traffic at the Isle of Vis. With the lower demand at Vis, men began to be reassigned to Zara or Bari as they were needed. During that first month Zara had 41 aircraft make emergency landings, and treated sixteen injured or wounded crew members, while Vis recorded 64 landings.

In March, Zara experienced so many landings that the squadron could not keep up with demand. Visits by Fifteenth Air Force generals resulted in attaching mechanics from the 451st, 454th, 455th, and 459th Bomb Groups, who arrived on 26 March 1945 and began repairing the planes that covered every parking space on the field. The same day Zara experienced the largest day in its history when 43 planes made emergency landings. For the month Zara took in 381 landings, 197,550 gallons of fuel and 4,800 gallons of oil were poured into aircraft and vehicles. The medical staff treated 42 combat wounded and 40 evacuees.

Over 35,000 meals were served. In contrast, Vis recorded 32 landings and Bari continued inspections, engine changes and general repairs.

During April 1945 and despite reduced targets, Zara still managed 168 landings including 4 crashes. Vis saw only one landing and spent most of their time in clean-up and salvage. Bari became busier due to the demand for cargo and troop carriers that accompanied the end of the war.

In May 1945 Zara's activity was limited to 8 P-38s that landed for fuel. There were no landings at Vis. The numbers told the story. For these emergency airfields the war was finally over. Throughout their existence the emergency fields and the 81st Air Service Squadron provided a safe haven for almost 1,000 aircraft of all types and thousands of crew members who otherwise would have been forced to risk the waters of the Adriatic Sea, saving countless American and allied lives. During December 1944 through April 1945 there were 60 landings by the 461st Bomb Group and 600 of their airmen. The Liberaiders were truly indebted to the 81st.

44-49674 of the 461st Bomb Group, 765th Bomb Squadron undergoing engine tests at Zara.

81st Air Service Squadron personnel repair battle damaged aircraft without the benefit of hangers regardless of the weather.

461st Bomb Group Aircraft Landings at Emergency Airfields

ISLE OF VIS

Date	S/N	Sq.	RCN	Nickname	Pilot	Sq.
Oct 1944	42-51346	766	31	Jake's Neighbor	R. J. Lauches	765
Oct 1944	42-51324	764	12	Ten Men Back/Paulette	E. A. Nahkunst	764
Oct 1944	44-41044	764	18	Lady Duzz	Brown	764
Dec 1944	42-51324	764	12	Ten Men Back/Paulette	C. B. Alexander	764
Dec 1944	44-41020	765	23	Judy R	L. L. Summers	765
Dec 1944	42-95304	766	42		J. M. O'Neal	766
Dec 1944	42-51918	766			G. F. Wiemann	766
Dec 1944	42-78676	764	2		C. A. Stevens	764
Dec 1944	42-51430	765	24	The Tulsamerican	E. P. Ford	765
Dec 1944	42-51894	764			M. C. Sargent	764
Dec 1944	42-51683	766	56		Frank	766
Dec 1944	42-52056	766	60		J. R. Yancey	765
Jan 1945	42-52056	766	60		H. L. Walsh	767
Jan 1945	42-51378	765	25		J. R. Yancey	765
Jan 1945	44-49619	767			J. H. Schultz	767
Jan 1945	42-52056	767	60		H. T. Walsh	767
Jan 1945	42-52052	765	27	Alma-M-Amur	C. H. BradyMaciejiwski	765
Jan 1945	42-52371	765	37	Upstairs Maid	T. Wright	765
Jan 1945	44-40269	766	55	Rough Rid'er	G. F. Weimann	766
Jan 1945	44-49428	765	29		J. E. White	765
Feb 1945	42-51894	765			R. A. Gress	764
Feb 1945	42-51695	767			C. W. Brady	765
Feb 1945	44-49879	Unknown			C. S. Alexander	764
Feb 1945	44-49952	767			L. D.Emmert	764
Feb 1945	42-51783	765	35	What's Next!	D. W. Michaelis	765
Feb 1945	44-49645	767	77		T. P. Willing	767
Feb 1945	44-49674	765	24	Maggie's Boomerang	D. Cameron	767
Feb 1945	42-51949	766	44		R. D. Faulkner	766
Mar 1945	42-78286	765	33		R. Chalmers	765
Mar 1945	42-51759	766			R. D. Faulkner	766

IESI

Date	S/N	Sq.	RCN	Nickname	Pilot	Sq.
Dec 1944	44-48796	767	77		J. W. Luebke	767
Dec 1944	42-95344	767	70		F. G. Stockton	767
Dec 1944	44-49586	767	64		R. E. Harrison	767
Dec 1944	42-51474	767	68		J. C. Kuestersteffen	767
Dec 1944	44-49641	767	63		G. R. Podwolski	767
Dec 1944	42-51838	767	65	Natural	W. M. Clarke	767
Jan 1945	42-51838	767	65	Natural	R. E. Harrison	767

ZARA

Date	S/N	Sq.	RCN	Nickname	Pilot	Sq.
Feb 1945	42-29564	767			M. A. Rush	767
Feb 1945	42-51918	766			C. T. Courtney	766
Mar 1945	42-50613	764	5	Jo	L. L. Crumbe	767
Mar 1945	41-29517	766	51	Pisces	R. L. Webb	766
Mar 1945	42-51474	765			K. B. Card	767
Mar 1945	42-51783	765	35	What's Next!	R. L. Brewster	765
Mar 1945	42-51867	765	27		E. B. Thomas	765
Mar 1945	42-52056	766	60		J. W. Boozer	764
Mar 1945	44-10557	765	27	You Bet!	R. J. Louches	765
Mar 1945	44-40269	766	55	Rough Rid'er	M. K. LaRock	766
Mar 1945	44-49619	Unknown			C. D. McGinnis	767
Mar 1945	44-49511	764	15	Miss Lace	W. E. Petty	764
Mar 1945	44-49588	Unknown			R. Missius	766
Mar 1945	44-49428	765	29		L. R. Heinze	765
Mar 1945	44-49623	767	78		R. J. Viau	767
Mar 1945	44-49674	765		Maggie's Boomerang	W. Miller	765
Mar 1945	44-49413	Unknown			J.C. Bontempo	766
Mar 1945	44-49887	764	6		E. R. Wolff	764
Mar 1945	44-49623	767	78		W.C. Owens	767

FALCONARA

Date	S/N	Sq.	RCN	Nickname	Pilot	Sq.
Nov 1944	44-48757	767	71		E. K. George	767
Dec 1944	42-51937	764	5		C. J. Schultz	764
Jan 1945	44-41112	765	26	Irish AngrIII/She Wolf	J. B. Moore	765
Feb 1945	41-29564	767			R. C. Hansen	767

ARMY AIRWAYS COMMUNICATIONS SERVICE: Station 158

THE ARMY AIRWAYS COMMUNICATIONS SERVICE (AACS) was established on 15 November 1938, based on a proposal by Lt. Col. Billy Mitchell, later the Chief of the Army Air Corps. At first its responsibilities included operating radio facilities improving radio traffic between airdromes across the United States. This would eventually require more than 700 detachments and would entail air-ground, ground-air, point-to-point communications pertaining to aircraft movement, controlling military air traffic, and navigational aid.

World War II brought about an entirely new set of responsibilities, including unified command, centralized flight control, and flexibility in the employment of tactical aircraft. It also supported the movement of ferry and transport aircraft as well as providing information that improved weather predictions. They were also responsible for coded messages and homing signals for aircraft navigation. Navigational aids might include Radio Ranges, Homing Beacons, or Loran.

Localizer

When an aircraft leaves the airfield he was cleared to leave the tower's frequency and tune into the radio range. A Radio Range transmitted two sets of signals, A and N, also known as dit-dah and dah-dit. The two tones were synchronized, resulting in a steady tone heard by the pilot. As long as the tone remains steady the pilot knew he was on one of the four legs of the range. However, when one of the tones became predominant, he knew he was off course and made the necessary corrections to get back on course. These ranges were aligned along a route, allowing the pilot to sign off one range and sign on to the new range. The final leg of the flight would have brought the pilot directly to the airfield where the tower would have given him verbal instructions to let down on the final approach. However, with other improvements in navigational aids, Radio Range became better suited as a pure navigational option.

"Even more strictly navigational aids than ranges are the various types of homing beacons used by the AACS". Transmitters broadcasted an omni-directional signal as a steady tone broken occasionally for a station identification, which usually could be heard at greater distances than the radio ranges. However, only aircraft with a radio compass installed could use them. When the radio compass was tuned to the homing signal, the pilot simply keeps the needle centered on the radio compass. As long as he heard a steady tone, and kept the needle centered, he was headed for his destination.

Instrument Approach System

In addition to navigational aids, AACS operated several landing systems which made it possible for an aircraft to land successfully with an almost zero ceiling and visibility.

The Instrument Approach System (IAS) was a mobile unit which could be set up at any runway at any airfield. Once in place and calibrated it sent a directional signal forming a glide path on the approach to the runway. Once the receiving equipment was installed on an aircraft, two needles controlled both lateral and vertical approach within a few yards of the runway.

Ground Control Approach (GCA) proved an even better system for guiding aircraft to a safe landing during bad weather. Unlike IAS, it did not require the installation of any equipment on the aircraft other than a standard radio for voice to voice communication. "GSA consists of a search system and a precision system operating on radar principles to provide a complete picture of the air traffic within a radius of 25 miles of the apparatus." The ground operators were able to keep a constant watch with their oscilloscopes which provided the exact location of the aircraft. The search system was used to follow and guide the aircraft until it reached the approach leg of the runway. At that point, the precision system took over. The operator gave the pilot exact instructions to get the aircraft on to the runway. The pilot repeated the instructions back to the operator as he made the corrections.

The GCA system also had redundant search and precision equipment. If anything went wrong while talking a pilot down, there was another system ready to take over.

Another option provided by AACS was high frequency (HF) and very high frequency (VHF) Direction Finding. Even with all the navigational aids it was possible for a pilot to get lost. The AACS had established directional finding nets made up of three or more stations in at least a 600 miles line. When a pilot wanted to know his location he simply called a station. He would send out a series of dashes and call letters which the stations used to take a bearing. Each station would triangulate their bearings to determine the exact location of the aircraft. Should the plane be damaged or short on fuel, they could also follow him for hours, keeping track of his exact location. If the plane is forced to ditch they could immediately contact sea-rescue. (See also "Big Fence")

The 18th Squadron was activated in North Carolina in October 1942 and assigned to the Twelfth Air Force. In March 1943 it was reassigned to the AACS African Airways Communications Area. It later moved along with the Twelfth Air Force to Italy the following November. As it was still responsible for African operations it was realized the 18th had become overextended. The 18th was divided into two sectors, with the European Sector headquartered in Naples. It also became responsible for eight new detachments. They bore the responsibility for providing navigational aid, aircraft control, and weather information for the Mediterranean Allied Air Force (MAAF), but had very limited support for the Twelfth and Fifteenth Air Forces.

The Naples Detachment of the 18th Squadron would have a short life, however. On 15 May 1944 AACS would completely reorganize into wings, groups, and squadrons. As a result, the 18th Squadron European Sector would be placed under the 2nd Wing in Casablanca. It would be renamed the 58th Group, headquartered in Naples, and would include the 116th and 117th Squadrons (Only the 116th would be activated on the 15th of May 1944). It would be men from the 116th Squadron that would eventually find their way to Torretta Field as Station 158.

Boundary Marker

On 3 February 1945, 2nd Lt. Edward Burks and six enlisted men received verbal orders to relocate from Tortorella to Torretta Field. Following vehicle and equipment checks they were ready to leave the next day with three jeeps, a localizer truck, and a glide path trailer. With Lt. Burks was Sgt. Carl S. Kolman (778), Sgt, Frank P. Mercore (754), Cpl. Claudius Hughes (033), Cpl. John M. Scowcroft (933), Cl, Charles A. Kalbfell (754), and Pfc N.P. Geannopoulos (778). They arrived at Torretta after a short drive and began setting up the new station immediately. Station 158, new Instrument Approach system went live on 6 February 1945 using identification call "TO". The new system was to be used by both the 461st and 484th. Training for the 461st was scheduled for February, with the 484th shortly thereafter. The 461st made two training flights before they cancelled the entire schedule. February would prove to be so busy

Sgt. Carl S. Kolman and Cpl. Claudius Hughes man the middle marker at Torretta Field in 1945

operationally that neither aircraft nor crew were available for training flights. Instead the month would be dedicated to installation and checking facilities. By mid-month, another eight enlisted men joined the station, including S/Sgt. Robert O. Monson (754), Sgt. Victor A. Supplee (754), Sgt. Clarence E. Lyden (754), Cpl. Irving H. Dunn (754), Cpl. Everette B. Davis (754), Cpl. J.A. Howenstein (933), Cpl. Joseph J. Valente (754), and Pfc. (754).

Operationally, March was no better than February. The equipment was used only twice by 461st cargo planes (converted B-24s) to test the installed equipment. Both tests were very satisfactory. The AACS was fully prepared to fulfill their mission at Torretta Field. All they lacked was customers.

In April, the 461st was still flying missions every day so there continued to be no customers to learn the equipment. However, the Fifteenth Air Force ordered the station to run its localizer beam from dawn to dusk every day to prepare pilots for flying bomb missions over various fronts. All five combat wings were assigned to practice flying the course during the month.

A glide path receiver was installed in a jeep which already had a localizer receiver allowing the jeep to check the glide path.

The group's last mission was flown on 26 April 1945. Although the 158th had been set up and ready to serve the needs of the 461st and 484th bombers for three months, they never did make a meaningful contribution to the war effort.

AACS Headquarters

There was little activity during May 1945, The localizer ran from 0800 to 1830 hours for the first nine days of May, but was shut down on orders from the Fifteenth Air Force on 10 May 1945. After that they did manage to provide some pilot training to the squadron, but this too was cancelled on the orders of the 461st commander.

The war was over for the two bomb groups at Torretta and it wasn't long before the base was closed. The need for navigational aids and communications, on the other hand, would continue to play an important role in post-war Europe. Station 158 would receive orders to pack up their equipment and head for their next assignment in Salzburg, Austria.

{6}

GOING HOME

REDEPLOYMENT AND DEMOBILIZATION

REDEPLOYMENT AND DEMOBILIZATION

IN A 14 JANUARY 1942 MESSAGE to Congress President Roosevelt gave the National Resources Planning Board (NRPB) responsibility for developing long-range plans for "post-war full employment, security, and building America." By mid-1942 the NRPB had conducted a conference with the War, Navy, and Labor Departments, War Manpower Commission, Veterans Administration, and several other agencies, resulting in a report, *Demobilization and Readjustment*, which was submitted to Roosevelt by 30 June 1942 and released to the public one month later. Planning would go through many refinements before a final plan was announced. In fact, final adjustments were still being made up to the end of the war in Europe. Considerations had to include:

- Planning had to be made with the understanding that two wars were being fought,
- Planners assumed the war in Europe would be won probably a year before the war against Japan, and
- Some part of the forces in Europe would have to reposition to fight the Japanese. It was further assumed the majority of those forces would come from the naval and air forces, with ground forces largely eligible for shipment home.

For some the end of the European war meant going home and getting on with life. For many, however, it meant heading for more fighting against Japan. It had been recognized from the beginning that winning a two-front war would be an extremely risky proposition. Therefore, the inherent philosophy for defeating the Axis would be concentrating on victory in Europe first, followed by pivoting forces to the Pacific after V-E Day to destroy Japanese troops. With that strategy in mind planning for the movement of forces in Europe and the Mediterranean began well over a year prior to the defeat of Germany. Aircraft, with the crews that had flown them, would be moved to the United States where facilities on the West Coast would process them for service against Japan. Other returnees would receive B-29 training before heading for combat.

The size of this effort would involve thousands of aircraft and tens of thousands of personnel. It would mean constructing facilities to handle the volume of people being moved around the world. Some groups, rather than going home themselves, would be reassigned to assist those who would be going. It would be broken into two parts, the GREEN PROJECT would bring home the men, while the WHITE PROJECT would bring home aircraft and their crews. Both would rely heavily on the Air Transport Command.

AIR TRANSPORT COMMAND

Air Transport Command, formerly Ferrying Command, was given "sweeping responsibilities" according to the official history of the Army Air Force. As ordered by the Commanding General, Army Air Forces, their responsibilities included the ferrying of all aircraft within the United States and to destinations outside the United States, control, operation, and maintenance of establishments and facilities on air routes outside of the United States, and transportation by air of personnel, materiel, and mail for all War Department facilities. Air Transport Command consisted of the Ferrying Division, which was responsible for moving aircraft overseas and ended up ferrying 267,000 aircraft to overseas locations by the end of the war, and Air Transport Division, which had the task of transporting primarily high value cargo and important personnel. For example, C-87s were used to deliver new engines to Libya to replace those worn out on the B-24s from the Ploesti mission.

By the end of the war Air Transport Command had grown into a huge and complex organization. Starting with 37,000 personnel it eventually grew to nearly 210,000 with 2,090 aircraft.

When Ferrying Command became Air Transport Command the South Atlantic sector became South Atlantic Wing, Air Transport Command. Personnel assigned to squadrons along that route would be assigned to a numbered station. Later, these numbered stations would be replaced by Army Air Force Base Units (AAFBU). For example, Dakar would become 1254th AAFBU and Marrakech would become 1257th AAFBU.

HOMEBOUND TASK FORCE

Col. E. M. Hampton, the Commanding Officer of the 463rd Bombardment Group (H), was placed in command of the Operation Homebound Task Force, which would plan and put into action the movement of personnel from Italy to Casablanca where they would be placed under the control of Air Transport Command. The 483rd Bombardment Group (H) would also be placed under his command.

Troops would be received from the 7th, 18th, and 27th Replacement Depots according to the following Schedule:

- 0 – 2,000 May 1945
- 5,000 June 1945
- 8,000 July 1945
- 12,000 August 1945
- 12 – 15,000 September & thereafter

After processing at Gioia these men were flown to Casablanca per the below assumptions:

1. 1. Fifteen passengers per aircraft
2. 2. 17 flights per day (255 passengers)
3. 3. 20 operational days per month (5,100 passengers)
4. 4. 42 aircraft would be assigned with 80% being operational
5. 5. 84 crews – 6 round trips per month

All personnel from this point would be either assigned to, or under the authority of, ATC for the purpose of movement to the United States. Once returned to the United States, control over the crews reverted to the Army Air Force. The Caribbean route through South America would carry approximately 50% of the passengers, the Atlantic route through the Azores would carry 40% more, and the Northern route was responsible for the remaining 10%.

GREEN PROJECT

The early strategy of the Allied forces was the defeat of Germany, after which the Allied forces would be transferred to the Pacific to defeat Japan. To accomplish this task ATC would receive an additional 33,000 people from combat bomb groups, theater troop-carrier groups, or their supporting organizations. They also received an additional 256 C-47 aircraft plus already planned C-54s. All of these troops would be transported with no reduction in any of their original responsibilities. As ATC would also be responsible for the care of all personnel under their control, they would also receive approval for requested new facilities in Casablanca, Port Lyautey, and Dakar.

The Green Project called for 50,000 passengers per month arriving home from Europe. To accomplish this goal the 460th Bomb Group was assigned to assist ATC. While scheduled to run until April 1946, the project was terminated on 10 September 1945 when shipping became available to transport most troops home.

There was a distinction between Green Project personnel and other normal traffic. It was assumed all Green Project personnel had been processed by the Air Force prior to turning them over to ATC. As a result they were handled differently from other passengers. There was no ticket; each man received a green tag that he wore on his uniform and another tag for his baggage. Even the instructions handed out by the ATC were printed on green paper. Once they arrived back in the United States they were promptly handed over to Army Service Forces.

One problem arose for which the ATC never found an answer. The combat groups had been led to believe, erroneously, that they would help the ATC for a few months and then go home as a proud unit. Instead they were absorbed into the ATC and later sent home individually. They had no time for serving those personnel they had seen as "Allergic to Combat." For their part, the ATC personnel saw no reason to give preferential treatment to a bunch of newcomers to their unit. An attempt was made to convince the unhappy groups that they did not work for the ATC, they were the ATC working for Uncle Sam. Nobody bought into the argument and morale remained low for the rest of the time the program existed. Despite their feelings about the situation, the former combat personnel did their jobs and did them well. The ATC, for their part, were handed a huge project without any reduction of their other assignments, including air evacuees and other "normal traffic" plus Eastbound traffic headed for India/China. That they accomplished all they set out to do is a testimony to their expertise and dedication.

Green Project

Target Month 1945	Total Passengers Westbound	Revised Target	Green Manifest	Air Evacuees	Other "Normal Traffic"	Total
May	16,000		695	7,468	12,219	20,382
June	27,500		13,649	5,876	10,589	30,114
July	36,000	43,760	37,704	6,127	6,683	50,514
August	50,600	50,600	36,682	5,578	5,714	47,047
1-15 Sept.	25,300	17,500	12,378	1,381	3,315	17,075
TOTAL			101,108	26,431	38,520	66,059

WHITE PROJECT

The White Project, developed in September 1944, was also under way, tasked with sending thousands of aircraft to America. It was decided that Gioia del Colle would be the point of departure for four-engine bombers and crews while medium bombers and two-engine fighters would leave from Pomigliano, Italy. Single engine aircraft would be processed for shipment by boat. Each ship would be flown by their combat crew plus a few passengers, but under the control of the ATC. Aircraft would be flown by the North Atlantic Route, Central Atlantic Route, and South Atlantic Route. Beginning in May 1945 25 aircraft per day would be flown home by the South Atlantic Route. A total of 2,282 aircraft and over 33,000 passengers came home on this route. April 1945 saw an end to combat aircraft arriving from the United States. First, four-engine bombers without radar were stopped. Then all bombers stopped and finally all aircraft stopped on April 18, 1945. Any planes still enroute to Europe on that date were turned around and sent back. The Combat Crew Reception Center at Gioia would not be welcoming any more men to Italy.

B-24s and B-17s stacked up at Gioia del Colle waiting for their flights to the Unites States as part of the White Program after the end of the war in Europe.

Once the missions to Spittal and Wolfsburg were completed, the 461st was assigned a role in the redeployment program by ferrying personnel. Their first task on 16 May 1945 was carrying members of the 484th Bomb Group to Gioia del Colle. Ferrying would continue throughout the week and would include trips to Pomigliano, Manduria, Amendola, Foggia, Fano, and Cattolica.

Following that assignment, air crew training began with an eye toward being ready for a long flight over water. Training included practice in ditching procedures, cross-country and night flying, instrument work, and celestial navigation. On 17 May 1945, the 461st received a phone message from HQ, 49th Bombardment Wing ordering the group to prepare for movement back to the United States on 4 June 1945. The Daily Report for Staging Airdrome No. 1, dated 29 May 1945 stated that the Fifteenth Air Force had been notified that ten B-24 aircraft with crews from the 461st Bomb Group were to arrive at Staging Airdrome No. 1 after 1600 hours on 29 May 1945, and another 31 B-24 aircraft with crews from the 461st were to arrive after 1300 hours on 30 May 1945. The training was put to work on 30 May 1945 when 31 aircraft and 31 crews departed Torretta Field enroute to Staging Airdrome #1 at Gioia del Colle for the first leg of the trip home. The following day another 25 ships with 45 crews followed the first echelon.

Gioia del Colle ceased to be a Fifteenth Air Force Service Command (XV AFSC) organization and came under the control of Allied Air Forces Service Command/Mediterranean Theater of Operations (AAFSC/MTO), which renamed it Staging Airdrome No. 1 under the direction of the 38th Air Depot Group. ("451st Bomb Group Newsletter No. 24 February 2024"). Pomigliano would be named Staging Airdrome No. 2 under the direction of the 515th Service Squadron. Any aircraft being processed for water shipment would go to Air Force General Depot Three in Naples, Italy under the direction of the 20th and 90th Air Depot Groups, which were composed of the 20th, 302nd, and 36th Depot Repair. Aircraft arrived at Gioia del Colle, the crew was picked up by truck, assigned temporary quarters including a cot and blankets, given showers, and sent to the mess hall. What happened from there was 36 hours of processing. As with the men who were traveling under the Green Project, it was assumed all out processing would be completed prior to being assigned to the ATC.

Clothing and flying equipment were checked and corrected as needed. Personnel forms were processed, including paybooks, flying records, shot cards, dog tags, medical forms, and other records.

Incoming crews were assigned temporary quarters while out processing

The pilot hands out forms to his crew prior to medical and dental processing

Crews receive US dollars in exchange for local currency

On the morning of departure bags containing flying gear and parachutes for each crewman and passenger are loaded on the aircraft

Medical and dental exams were conducted. Just prior to leaving, local currency was exchanged for U.S. currency. During this processing period extra personnel were assigned to the plane as passengers. The aircraft was also being put through its paces. All information pertaining to the plane was handed over to the Alert Section by the pilot. Following review of the documentation a team of mechanics went over the entire aircraft fixing anything found no matter how small. Once cleared for flight testing it was turned over to the Flight Section for a thorough flight test by depot-based pilots. If satisfactory, the engineering liaison officer notified the base personnel processing officer. The crew was alerted and ordered to fly a two-hour fuel consumption test.

The next morning the crew received a two-hour briefing and was cleared for takeoff. During the flight from Gioia to Marrakech the aircraft and crew remained under the operational control of the theater, but would transfer to the control of the ATC once they arrived in Marrakech. This process was repeated 25 times every day, seven days a week.

From Marrakech the planes flew all three routes homes, including 3,224 by the Northern route, 2,282 by the South Atlantic route and another 459 heavy bombers by the Middle Atlantic route through the Azores. In addition to the 50,763 crewmen who flew home there were an additional 33,850 passengers. In all the White Project brought over 84,000 men home.

The Green Project and White Project were designed to quickly move the aircraft and crews that would be needed for the final victory over Japan. The ATC moved thousands of aircraft and hundreds of thousands of personnel. However, with the sudden surrender of Japan, not one bomber or crew ever went to the Pacific Theater.

The aircraft that came home, intended for service in the Pacific Theater, were no longer needed. Thousands of B-24s and B-17s ended their days at bone yards around the country, many at Walnut Ridge, Arkansas. There they waited, row after row, for their turn in the smelters to be turned into bars of aluminum. Gone were the days they carried ten brave men into combat and brought them home again. Their engines would no longer roar, the guns no longer spit fire at the enemy, their crews no longer fighting for a worthy cause. Left to their memories, they sat quietly awaiting a fate none of them could avoid.

Taken 15 Nov 1945, B-24s and B-17s await their turn being broken up and melted down at the Reconstruction Finance Corporation Storage, Disposal, and Salvage Facility in Walnut Ridge, Arkansas

GROUND ECHELON

While the air echelon was returning to the Zone of the Interior, the ground echelon would be headed home by ship. A phone call from the 49th Bombardment Wing on 17 May 1945 ordered the 461st to prepare for movement on 4 June.

The first step was clearing the area, and preparing all records, equipment, and personnel for shipment. However, three days before departure, the order was cancelled. As everything was already prepared for movement, all the men could do was wait for further orders. The air crews had begun leaving for Staging Airdrome No. 1 at Gioia del Colle by 25 May and were all on the way home by 20 June.

The ground echelon remained at Torretta Field until 28 June when the Fifteenth Air Force scheduled their movement to PBS Staging Area #1 at Naples. The orders that arrived set their time of arrival at Naples on 7 July. Almost immediately, the travel schedule was moved forward to allow for arrival At Naples on 1 July. As the group had been ready for some time, there were no problems with the earlier departure date. Although the original orders called for a shipment of 1,300 men and officers, only around 1,000 remained together in the group.

The Special Planning Division had been assigned the task of devising a system to determine who would occupy Germany, who would transfer to the Pacific, and who would be sent home. Following the end of World War I, the US Army's rapid and disorganized demobilization left it entirety too undermanned. After World War I, moreover, American divisions had been sent home as entire units. During World War II this was not an option since older units included many crews who had arrived overseas as replacements in the final months of the war.

The Army had to find a way to send soldiers home in a prompt manner based on objective criteria. Otherwise it risked provoking widespread protests from restless soldiers overseas who had no enemy to fight.

After soliciting feedback from commanders around the world, the Army devised and implemented a system called the Adjusted Service Rating Score, more commonly referred to it as the "point" system. Under point system, everyone was awarded a number of points based on how long they had been overseas, how many decorations they had received, how many campaigns they had taken part in, and how many children they had. ("The Points Were All That Mattered: The US Army's Demobilization After ...")

Points were awarded according to the following formula:

- One point for each month in service in the Army
- One additional point for each month in service overseas
- Five points for each campaign
- Five points for a medal for merit or valor
- Five points for a purple heart (awarded to all soldiers who were wounded in action)
- Twelve points for each dependent child up to three children

USS Mount Vernon

First, enlisted men with over 100 points, and later those with over 85 points were transferred from the group before it boarded the transport for the United States. This was necessary because the group was committed to redeployment at that time, and these "high point" men were to wait discharge.

Embarkation of the 461st Bomb Group began on 8 July. In the morning, the group began boarding the transport USS Mount Vernon, and by 1800 hours was under way back home. Ten days later the transport arrived at Hampton Roads, Virginia. Entertainment and excellent food made the trip far more enjoyable than the original one-month trip to Italy.

On 19 July 1945, the officers and men in the group made their way to various reception centers where they were processed and sent on 30-day furloughs and leaves.

On 25 August 1945, the 461st men began arriving at Sioux Falls Army Air Field. Two days later, on 27 August, the group was inactivated in accordance with G.O. 116, Section II, paragraph 1, Headquarters, Second Air Force, Colorado Springs, Colorado, dated 27 August 1945. The personnel were sent either to separation centers, or to other bases, mostly in the Second Air Force.

The group had been awarded two War Department Unit Citations. The first citation was for the attack on the Duna Repulepgyar Aircraft Factory at Budapest. Hungary, on 13 April 1944. The cluster to the citation was awarded for an instrument bombing mission against the Creditul-Minier Oil Refinery, Ploesti, Romania on 15 July 1944. Individual decorations of group members included: 3 awards of the Legion of Merit, 10 Silver Stars, 264 Distinguished Flying Crosses, 27 Oak Leaf Clusters to the DFC, 12 Soldier's Medals, 58 Bronze Stars, 2,619 Air Medals, 3,709 Oak Leaf Clusters to the Air Medals, 261 Purple Hearts, and 8 Oak Leaf Clusters to the Purple Hearts.

The group flew 223 combat missions, dropping 10,885 tons of bombs on target. 461st gunners destroyed 129 enemy fighters, probably destroyed 44 more, and damaged 16 others. They lost 108 bombers, including 45 to anti-aircraft, 32 to enemy fighters, and 31 to other combat losses. The average monthly number of aircraft assigned was 56.9, while there was a monthly average of 83.9 crews.

Made in the USA
Monee, IL
05 November 2026

67810992R00167